PRINTING INK AND
OVERPRINT VARNISH FORMULATIONS

PRINTING INK AND OVERPRINT VARNISH FORMULATIONS

Second Edition

by

Ernest W. Flick

np

NOYES PUBLICATIONS
WILLIAM ANDREW PUBLISHING, LLC

Norwich, New York, U.S.A.

Library of Congress Catalog Card Number: 99-20463
ISBN: 0-8155-1440-9
Printed in the United States

Printed and bound in the United Kingdom

Transferred to Digital Printing, 2011

Library of Congress Cataloging-in-Publication Data

Flick, Ernest W.
Printing ink and overprint varnish formulations / by Ernest W. Flick. -- 2nd ed.
p. cm.
Includes index.
ISBN 0-8155-1440-9
1. Printing-ink. I. Title.
TP949.F57 1999 99-20463
667'.5--dc21 CIP

In Memory of My Father

Preface

This is a collection of more than 200 formulations for printing inks, overprint varnishes, and related products. The data represent selections from manufacturers' descriptions made at no cost to, nor influence from, the makers or distributors of these materials. Only the most recent formulas have been included. It is believed that all of the trademarked raw materials listed here are currently available, which will be of utmost interest to readers concerned with raw material discontinuances.

The formulations in this book are divided into the following 4 sections:

Section I Printing Inks
Section II Overprint Varnishes
Section III Coatings
Section IV Miscellaneous

In addition to the above, there are two other sections which will be helpful to the reader:

Section V A chemical trademark section where most tradenamed raw materials included in the book are listed with a chemical description and the supplier's name. The specification which each raw material meets are included, if applicable.

Section VI Main office addresses and phone numbers, including 800 numbers, of the suppliers of trademarked raw materials, some of which are not available in the usual reference books.

The book lists the following product information, as available, in the manufacturer's own words:

1. Description of end use.

2. The formulas include the following:
   A. A listing of each raw material contained.
   B. The amount of each raw material included in the formula, as presented by the supplier.
   C. An description of the formula.
   D. The formula source, which is the company or organization, that supplied the formula. A secondary source may be the originating company and/or the primary source's publication title, or both. A formula number is included, if applicable.

An End Use Index has also been included in the book for rapid location of the various types of ink and varnish formulations.

May 1999 Ernest W. Flick

NOTICE

To the best of our knowledge the information in this publication is accurate; however, the Publisher does not assume any responsibility for the accuracy or completeness of, or consequences arising from, such information. This industrial guide does not purport to contain detailed user instructions, and by its range and scope could not possibly do so. Mention of trade names or commercial products does not constitute endorsement or recommendation for use by the Publisher.

In some cases ink and varnish formulations may contain potentially toxic materials, and therefore due caution should be exercised and expert advice obtained when handling these materials. Final determination of the suitability of any information or product for use contemplated by any user, and the manner of that use, is the sole responsibility of the user. We strongly recommend that users seek and adhere to a manufacturer's or supplier's current instructions for handling each material they use.

The Author and Publisher have used their best efforts to include only the most recent data available. The reader is cautioned to consult the supplier in case of questions regarding current availability.

Contents and Subject Index

Section I
Printing Inks

Adhesion Promoters

Titanium Acetyl Acetonate (TAA) Derivatives:

The use of TAA derivatives in a CAP-based ink formulation containing a high-acid-number maleic or acrylic resin was found to improve adhesion to coextruded films such that commercial inks were introduced in the late '70s. The use of modifying resins having apppreciable acid values was necessary in the formulation to obtain adequate adhesion to the coextruded polyolefin substrate.

Table 4 shows a typical ink formulation. The adhesion promoter (P1-2) in the formulation is isopropoxyethoxy titanium bis acetylacetonate, which was preferred over the unmodified TAA since it does not crystallise in cold weather. The titanium derivative is thought to promote adhesion by reacting with hydroxyl groups on the CAP resin and with groups introduced into the polyolefin film during the corona discharge treatment.

Table 4

CAP/Acrylic Ink for Coextruded Polyolefin Films

Ingredients:	Wt%
CAP-482-0.5	8.6
Surcol 836W acrylic resin	4.3
Tioxide R-TC 90 titanium dioxide	25.8
Dibutyl phthalate	1.7
Tilcon P1-2 (65%) isopropoxyethoxy titanium bis acetylacetonate	2.0
Lancoglidd TD (25%) wax	2.0
1:1 denatured ethanol:ethyl acetate, 99%	55.6

Notes:

*This ink may be overprinted with an unmodified CAP/acrylic coloured ink as shown in Table 2.

*The drying rate may be retarded, if necessary, by adding 1%-5% n-Butyl acetate.

*CAP-504-0.2 may be substituted for CAP-482-0.5 if a greater concentration of ethanol in the solvent blend is required.

The ink stability, which appears to be in excess of 12 months, depends on both the presence of alcohol in the solvent blend and on the acid value of the modifying resin. The acid value of the modifying resin in Table 4 is around 60; if resins having higher values are used, the ink stability is impaired.

SOURCE: Eastman Chemical Co.: Formulation W16

Black Ink

Raw Materials:	Wt%
Place into clean ball mill for 12 hours.	
Carboset GA-1364	69.2
DM Water	18.0
Monoethanolamine	0.3
Foamburst 151	0.4
Foamburst 157	0.1
Raven 450	12.0
Check grind.	

Formulation Constants:
Total Solids by Weight, %: 30.0
Total Solids by Volume, %: 23.0
pH: 8.7
Weight per Gallon/lbs: 9.1
Viscosity, Brookfield, 25C, cps: 211
Viscosity, #3 Zahn, 25C, sec.: 12
Formulation WI 62

Red Ink

Raw Materials:	Wt%
Place into clean ball mill for 12 hours.	
Carboset GA-1364	69.2
DM Water	18.0
Monoethanolamine	0.3
Foamburst 151	0.4
Foamburst 157	0.1
Sunbrite Red	12.0
Check grind.	

Formulation Constants:
Total Solids by Weight, %: 30.0
Total Solids by Volume, %: 23.0
pH: 8.8
Weight per Gallon/lbs.: 9.0
Viscosity, Brookfield, 25C, cps: 443
Viscosity, #3 Zahn, 25C, sec: 16
Formulation WI 159

SOURCE: BF Goodrich Specialty Chemicals: Carboset Formulations

Black Ink for Paper Board

Raw Materials:	Wt%
Place into a clean mixing vessel:	
Liquaflex Black	34.0
Slowly mix and add:	
Carboset GA-1086	35.0
CL 119 (See page 92)	18.0
Michem 32535	3.0
CL 135 (See page 92)	3.4
Water	4.3
Butyl Cellosolve	2.2
BYK 301	0.1
Mix at medium speed for 30 minutes	

Formulation Constants:
- Total Solids by Weight, %: 41
- Total Solids by Volume, %: 36
- pH: 8.1
- Weight per Gallon, lbs: 9.1
- Viscosity, #3 Zahn, 25C, sec.: 20
- Viscosity, Brookfield, 25C, cps: 135

Formulation WI 141

Overprint Varnish

Raw Materials:	Wt%
Place into a clean mixing vessel:	
Carboset GA-1086	60.0
Slowly mix and add:	
CL 119 (See Page 92)	29.1
Michem 32535	2.8
CL 135 (See Page 92)	2.0
Water	4.0
Butyl Cellosolve	2.0
BYK 301	0.1
Mix at medium speed 30 minutes	

Formulation Constants:
- Total Solids by Weight, %: 43
- Total Solids by Volume, %: 40
- pH: 8.1
- Weight per Gallon, lbs: 8.7
- Viscosity, #3, 25C, sec.: 20
- Viscosity, Brookfield, 25C, cps: 160

Formulation CL 142

SOURCE: BF Goodrich Specialty Chemicals: Carboset Formulations

Blue Ink

Raw Materials:	Wt%
Flexiverse BFD-1121	30.0
Mix and add slowly while mixing:	
CL 106B (See below)	20.0
Water	11.0
Carboset GA-1087	31.0
Santicizer 160	3.5
Michem 32535	4.5

Mix at medium speed for 15 minutes.

Formulation Constants:
Total Solids by Weight, %: 44
Total Solids by Volume, %: 39
pH: 8.8
Viscosity, 25C, #3 Zahn, Seconds: 36
Weight per Gallon, lbs: 9.1
Formulation WI 131

High Solids Resin Cut

Raw Materials:	Wt%
Place into a clean mixing vessel:	
Water	27.0
Isopropanol	10.0
Ammonia, 28%	13.0
Add slowly while mixing:	
Carboset GA-1161	50.0

Mix at high speed for 45 minutes or until all resin has dissolved

Formulation Constants:
Total Solids Weight, %: 52.0
Total Solids by Volume, %: 47.6
pH: 8.2
Viscosity, Brookfield, 25C, cps: 3,250
Weight per Gallon, lbs: 8.9
Formula CL 106B

SOURCE: BF Goodrich Specialty Chemicals: Carboset Formulations

Blue Ink for Paper and Paper Board

Raw Materials:	Wt%
Place into a clean ball mill:	
Carboset GA-1387	65.0
DM Water	16.8
Isopropyl Alcohol	3.0
BYK 035	0.2
Sunfast Blue 249-1291	15.0
Ball mill 12 hours - check grind	

Formulation Constants:
- Total Solids by Weight, %: 30.0
- Pigment binder ratio: 1:1
- Viscosity, #2 Zahn, 25C, sec: 22
- pH: 8.2

Formulation WI 138

Blue Corrugated Ink

Raw Materials:	Wt%
Slowly Premix the following:	
Sunfast Blue 449-1281	30.0
Carboset GA-1387	60.0
Isopropyl Alcohol	3.0
Huber 900	5.0
Foamburst 338	0.2
Water	1.1
Ethanolamine	0.7

Place into a clean ball mill and ball mill 12 hours - check grind.

Formulation Constants:
- Total Solids by Weight, %: 30.8
- Total Solids by Volume, %: 23.6
- pH: 9.0
- Weight per gallon, lbs.: 9.2
- Viscosity, #2 Zahn, 25C, sec.: 29
- Viscosity, Brookfield, 25C, cps: 135
- Pigment to Binder Ratio: 1/1.2

Formulation WI 140

SOURCE: BF Goodrich Specialty Chemicals: Carboset Formulations

Bronze Gold Ink

Components:	Wt%
Eastek 2140 polymer dispersion (a)	47.2
Eastek 1300 polymer dispersion (a)	11.7
n-Propyl alcohol	5.0
Surfynol 104 PA surfactant (b)	0.5
Cobratec 99 benzotriazole (c)	0.2
Defoamer (d)	0.1
Associative thickener (e)	0.2
Biocide (f)	0.1
Richgold flake 9520X (bronze powder) (g)	<35.0

(a) Eastman
(b) Air Products
(c) PMC Specialties
(d) Suggested defoamers include DC 65 Additive (Dow Corning); Surfynol DF-75 (Air Products); Dee Fo 97-3 (Ultra Additives); Defoamer RD-1386 (R&A Auxiliaries); Mazu DF 327C (PPG).
(e) Suggested associative thickeners include Acrysol RM-825 (Rohm and Haas); DSX-1514 (Henkel).
(f) Suggested biocides include Tektamer 38LV (Calgon); Proxel GXL and Proxel BD (Zeneca).
(g) U.S. Bronze Powders

Formulation Notes:
1. Components are listed in order of addition.
2. Ink viscosity will depend on bronze level. To adjust viscosity, this formulation can be cut with water or with an 80/20 blend of water/n-propyl alcohol.

Application: Gravure

Substrates:
Gift wrap
Other coated papers

Features:
*Good metallic appearance
*Redispersible after settling
*Fast running speeds

SOURCE: Eastman Chemical Co.: Eastek Standard No. 91501

CAP/Acrylic Ink With Amino Silane Adhesion Promoter

Silanes:

Silanes react with hydroxyl groups on CAP and promote adhesion similarly to the titanium derivatives described. However, it has been found that amino silanes have the greatest effect on adhesion improvement. The amino silane derivatives that appeared to offer the best properties were Silane Z6020 from Dow Corning or Dynasilan DAMO from Dynamit Nobel.

The length of the alkyl chain on the amino silane appeared to have little effect on adhesion, but the longer-chain silanes produced a strong amine odour during ink drying. Another aspect of using amino silanes is that formulated inks cannot be successfully printed immediately. There appears to be some chemical reaction between the silane and the binder; if printing is carried out before it is completed, poor adhesion results.

With Silane Z6020 the delay was 16 hours, but with new samples of Dynasilan DAMO the delay was 10 minutes, although it increased with aging. The delay, however, could be obviated by preparing master solutions of silane with acrylic resin solution. These appeared stable to long-term aging and could be used in inks for immediate use.

Table 7 shows a typical CAP-based silane ink formulation. For the silanes examined, it was important that the acrylic resin has a zero acid number; acrylics having a measurable acid value resulted in poor gloss. One disadvantage of silane-based inks was their lack of viscosity stability during aging over several weeks.

Table 7

Ingredients:	Wt%
CAP-482-0.5	8.73
Plexigum N742 acrylic resin	4.37
Dibutyl phthalate	1.75
Tioxide R-TC 90 titanium dioxide	26.20
Silane Z6020 aminosilane derivative	1.31
1:1 industrial ethanol:ethyl acetate	57.64

The viscosity increase, however, could be reduced somewhat by adding 2.6% trimethylorthoformate or triethylorthoformate to the formulation in Table 7. The use of Dynasilan 1411 aminosilane derivative, a chemically modified Dynasilan DAMO, resulted in a reduced viscosity increase during ink aging.

The resulting CAP-based ink print properties appeared good and compared quite favourably with the unmodifed amino silane derivative. As with the polyethylene imine adhesion promoter, silanes could not be used with nitrocellulose-based inks; gelation occurred almost immediately.

SOURCE: Eastman Chemical Co.: Formulation W44

CAP/Acrylic Ink for PVDC-Coated Films

In Europe probably the most common films printed with CAP-based inks are PVDC-coated cellophane or polypropylene. Acrylics are the most popular modifying resins for these CAP-based inks because they offer a combination of essential properties.

The most important properties are low odour and low retained solvent, high heat resistance, the ability to be overlacquered with PVDC, and excellent adhesion. Table 2 shows a typical formulation of such an ink. The acrylic resin and plasticiser in the formulation may be altered to suit printing conditions.

In the formulation in Table 2, the white base ink consists of 2:1 CAP:acrylic resin to produce adequate heat resistance. The pigment:binder ratio is 2:1 in order to provide hiding. The top ink contains 3:1 CAP:acrylic resin to improve the heat resistance as much as possible and to reduce the odour, since acrylic polymers may contain residual monomer.

Table 2
CAP/Acrylic Ink for PVDC-Coated Films
Base Ink

Raw Materials:	Wt%
CAP-482-0.5	8.6
Paraloid B72 resin	4.3
Tioxide R-TC 90 titanium dioxide	25.8
Dibutyl phthalate	2.0
Lancoglidd TD (25%) wax	4.5
Crodamine OR wax	1.5
1:1 denatured ethanol:ethyl acetate	53.3

Top Ink

Raw Materials:	Wt%
CAP-504-0.2	9.0
Plexigum N80 resin	3.0
Dibutyl phthalate	2.0
Pigment	12.0
Lancoglidd TD (25%) wax	2.3
4:1 denatured ethanol:ethyl acetate	71.7

Notes:

*It was found that in those cases where the drying rate was too rapid the addition of 1%-5% n-butyl acetate was beneficial.

*The above formulation may also be used on acrylic-coated polypropylene (e.g. Bicor MB600) or treated polyester film (e.g. Melinex 813).

SOURCE: Eastman Chemical Co.: Suggested Formulations

CAP/Acrylic Ink with Zirconium Propionate

Zirconium Derivatives:

Several zirconium compounds act as adhesion promoters for CAP based inks, probably by reacting with available hydroxyl groups on the CAP molecule similarly to the previously described titanium derivatives. Zirconium propionate exhibits the correct solubility, compatibility, and reactivity to be useful with cellulose acetate propionate and was formulated into printing inks as Table 8 shows.

Table 8

Ingredient:	Wt%
CAP-482-0.5	8.73
Plexigum N742 acryic resin	4.37
Dibutyl phthalate	1.75
Tioxide R-TC 90 titanium dioxide	26.20
Zirconium propionate	1.31
1:1 industrial ethanol:ethyl acetate	57.64

Notes:

*This ink may be overprinted with an unmodified CAP/acrylic coloured ink as Table 2 shows.
*The drying rate may be retarded, if necessary, by adding 1%-5% n-butyl acetate.
*CAP-504-0.2 may be substituted for CAP-482-0.5 if a greater concentration of ethanol in the solvent system is required.

As with the silane adhesion promoters, the use of a zero-acid-number acrylic resin was necessary to produce the best gloss, compatibility, and shelf life. Zirconium propionate-based formulations did not appear to increase in viscosity with time, and useful shelf life appeared to be in excess of 12 months. The formulations also adhered successfully on metallised propylene film. The adhesion improving markedly with overnight aging.

SOURCE: Eastman Chemical Co.: Formulation W41

Coalesced Joncryl 537

Joncryl 537 is an acrylic emulsion that has been specifically designed for its alkali resistance properties. It is a hard polymer that needs to be coalesced (softened) in order to maximize its resistance and film forming properties.

	% by Weight
Joncryl 537	95.0
Glycol Ether: Butyl Cellosolve & Ektasolve EB	5.0

Procedure:
Add the coalescing solvent to the Joncryl 537 with moderate agitation. Continue agitation until homogeneous.

Alkali Resistant Ink-Blue

	Wt%
Pigment Blue: Sun Fast-248-1282	15.0
Joncryl 56	28.0
Coalesced Joncryl 537	26.0
Water	24.0
Jonwax 26	7.0

Viscosity (#2 Zahn): 20 secs

Alkali Resistant Ink-Red

	Wt%
Pigment Red: Sun Fast - 234-0077	15.0
Joncryl 56	28.0
Coalesced Joncryl 537	26.0
Water	24.0
Jonwax 26	7.0

Viscosity (#2 Zahn): 22 secs

Alkali Resistant Ink-Yellow

	Wt%
Pigment Yellow: Sun Brite - 273-3556	15.0
Joncryl 56	28.0
Coalesced Joncryl 537	26.0
Water	24.0
Jonwax 26	7.0

Viscosity (#2 Zahn): 21 secs

SOURCE: SCJohnson Polymer: Joncryl 56 Suggested Formulations

Diarylide Yellow Ink for Polyethylene/Board

Raw Materials:	Wt%
Place into a clean mixing vessel:	
Liquaflex Yellow YD-3000	26.8
Slowly mix and add:	
Carboset GA-1324	57.8
Water	5.0
n-Propyl Alcohol	3.0
Michem 32535	7.0
Foamburst 338	0.1
BYK 301	0.2
Fluorad FC120	0.1
Mix at medium speed for 15 minutes	

Total Solids by Weight, %: 42.7
pH: 8.7
Weight per Gallon, pounds: 8.8
Viscosity, Brookfield, 25C, cps: 94

P.E. Film:
Gloss, 60C: 106
Coefficient of friction (COF): 88.2

Carolina Board:
Gloss, 60C: 35
Blocking, Face-to-Face, F: 280
Slide Angle: 33.6
Coeffient of Friction (COF): 88.0

Formulation WI 136

Oil Resistant OPV

Raw Materials:	Wt%
Place into a clean mixing vessel:	
Carboset GA-1324	50.0
Premix and add slowly while mixing:	
Water	20.3
Isopropyl Alcohol	4.0
Carboset GA-1931	20.8
Michem 32535	4.4
Mix at moderate speed for 30 minutes.	
Add slowly while mixing.	
BYK 301	0.3
FC 129	0.2
Agitate at moderate speed until thoroughly mixed.	

Formulation Constants:
Total Solids by Weight, %: 36.0
Total Solids by Volume, %: 34.3
pH: 8.6
Weight per Gallon, pounds: 8.7
Viscosity, Zahn #2, seconds: 26.0
Viscosity, Brookfield, 25C, cps: 54

Formulation CL-164

SOURCE: BF Goodrich Specialty Chemicals: Carboset Formulations

Extender for Beverage Board Inks

Components:	Wt%
Eastek 2140 polymer dispersion (31321) (a)	74.5
Eastek 1300 polymer dispersion (31281) (a)	18.7
Surfynol 104 PA surfactant (b)	0.5
n-Propyl alcohol	6.0
Acrysol RM-825 associative thickener (c)	0.2
Biocide (d)	0.1

(a) Eastman
(b) Air Products
(c) Rohm and Haas
(d) Suggested biocides include Tektamer 38LV (Calgon); Proxel GXL and Proxel BD (Zeneca).

Formulation Notes:

1. Components are listed in order of addition.
2. Add pigment color bases to the extender to achieve desired shade. Cut with 50/50 water/alcohol to printing viscosity (typically 20-25 s/#2 Zahn EZ series cup). At press-side, add more extender or pigment color base as necessary to adjust color strength.

Application: Gravure
Substrate: Beverage board
Features:
*Fast running speeds
*User-friendly on press
*Compatible with many pigment dispersions and mill bases
*High gloss

SOURCE: Eastman Chemical: Eastek Standard No. 91712

Blue Ink for Polyolefins

Material:	Wt%
Place into clean mixing vessel:	
Carboset XPD-1993	47.0
Add while mixing:	
Isopropyl Alcohol	4.0
Trionic 100	2.5
Surfynol 104PA	0.5
Flexiverse BFD-121	43.0
Water	3.0

Formulation Constants:
Total Solids by Weight, %: 42.4
pH: 8.7
Weight per Gallon, pounds: 8.9
Viscosity, Zahn #2, 25C, seconds: 39.0
Print viscosity: 25.0

SOURCE: BF Goodrich: Formulation WL-240

Finished Inks

Pigment dispersions made from Joncryl 67 (see page 100) provide a good basis on which to formulate water based fluid inks from Joncryl 74.

Quality Corrugated-Red

	Wt%
Dispersion A	43.0
Joncryl 74	11.0
Joncryl 87	39.0
Jonwax 26	5.0
Water	1.5
Antifoam: Bubble Breaker 748	0.5

Initial Viscosity*: 25
One Week Viscosity*:
Before agitation: 35
After agitation: 31
*Secs. #2 Zahn Cup

Quality Corrugated-Blue

	Wt%
Dispersion B	37.0
Joncryl 74	11.0
Joncryl 87	42.0
Jonwax 26	5.0
Water	4.5
Antifoam: Bubble Breaker 748	0.5

Initial Viscosity*: 25
One Week Viscosity*:
Before agitation: 28
After agitation: 27
*Secs, #2 Zahn Cup

Quality Corrugated-Yellow

	Wt%
Dispersion C	37.0
Joncryl 74	11.0
Joncryl 87	42.0
Jonwax 26	5.0
Water	4.5
Antifoam: Bubble Breaker 748	0.5

Initial Viscosity*: 26
One Week Viscosity*:
Before agitation: 36
After agitation: 32
*Secs, #2 Zahn Cup

SOURCE: SC Johnson Polymers: Joncryl 74 Suggested Formulations

Finished Inks

Pigment dispersions made from Joncryl 67 (see page 100) provide a good basis on which to formulate water based fluid inks from Joncryl 74.

Economy Corrugated-Red

	Wt%
Dispersion A	43.0
Joncryl 74	10.0
Joncryl 91	18.5
Jonwax 22	5.0
Water	23.0
Antifoam: Bubble Breaker 748	0.5

Initial Viscosity*: 26
One Week Viscosity*:
Before agitation: 32
After agitation: 29
*Secs, #2 Zahn Cup

Economy Corrugated-Blue

	Wt%
Dispersion B	37.0
Joncryl 74	10.0
Joncryl 91	20.0
Jonwax 22	5.0
Water	27.5
Antifoam: Bubble Breaker 748	0.5

Initial Viscosity*: 26
One Week Viscosity*:
Before agitation: 29
After agitation: 28
*Secs, #2 Zahn Cup

Economy Corrugated-Yellow

	Wt%
Dispersion C	37.0
Joncryl 74	10.0
Joncryl 91	21.0
Jonwax 22	5.0
Water	26.5
Antifoam: Bubble Breaker 748	0.5

Initial Viscosity*: 25
One Week Viscosity*:
Before agitation: 22
After agitation: 21
*Secs, #2 Zahn Cup

SOURCE: SC Johnson Polymers: Joncryl 74 Suggested Formulations

Finished Inks

Pigment dispersions made from Joncryl 67 (see page 100) provide a good basis on which to formulate water based fluid inks from Joncryl 74.

Foil, Film, Metallized, & Tissue-Red

	Wt%
Dispersion A	43.0
Joncryl 74	51.5
Jonwax 26	5.0
Antifoam: Bubble Breaker 748	0.5

Initial Viscosity*: 25
One Week Viscosity*:
Before agitation: 39
After agitation: 33
*Secs, #2 Zahn Cup

Foil, Film, Metallized, & Tissue-Blue

	Wt%
Dispersion B	36.0
Joncryl 74	53.5
Jonwax 26	5.0
Water	5.0
Antifoam: Bubble Breaker 748	0.5

Initial Viscosity*: 26
One Week Viscosity*:
Before agitation: 28
After agitation: 27
*Secs, #2 Zahn Cup

Foil, Film, Metallized, & Tissue-Yellow

	Wt%
Dispersion C	37.0
Joncryl 74	51.5
Jonwax 26	5.0
Water	6.0
Antifoam: Bubble Breaker 748	0.5

Initial Viscosity*: 25
One Week Viscosity*:
Before agitation: 29
After agitation: 26
*Secs, #2 Zahn Cup

SOURCE: SC Johnson Polymers: Joncryl 74 Suggested Formulations

Finished Inks

Pigment dispersions made from Joncryl 67 (see page 100) provide a good basis on which to formulate water based fluid inks from Joncryl 74.

Tag & Label-Red

	Wt%
Dispersion A	43.0
Joncryl 74	22.0
Joncryl 87	27.0
Jonwax 26	5.0
Water	2.5
Antifoam: Bubble Breaker 748	0.5

Initial Viscosity*: 25
One Week Viscosity*:
Before agitation: 40
After agitation: 30
*Secs, #2 Zahn Cup

Tag & Label-Blue

	Wt%
Dispersion B	37.0
Joncryl 74	23.5
Joncryl 87	30.0
Joncryl 26	5.0
Water	4.0
Antifoam: Bubble Breaker 748	0.5

Initial Viscosity*: 25
One Week Viscosity*:
Before agitation: 29
After agitation: 28
*Secs, #2 Zahn Cup

Tag & Label-Yellow

	Wt%
Dispersion C	37.0
Joncryl 74	23.0
Joncryl 87	29.5
Jonwax 26	5.0
Water	5.0
Antifoam: Bubble Breaker 748	0.5

Initial Viscosity*: 25
One Week Viscosity*:
Before agitation: 43
After agitation: 33
*Secs, #2 Zahn Cup

SOURCE: SCJohnson Polymer: Joncryl 74 Suggested Formulations

Flexographic Inks

The following formulations are recommended as a starting point for flexographic inks to be used in surface printing of treated polyethylene. Modification may be necessary for individual applications.

Nitrocellulose Modification-HM Ink

	Wt%
Day-Glo HM Soluble Toner	35.0
1/4 Sec. SS Nitrocellulose (70% dry basis in Ethanol)	10.8
Anhydrous Ethyl Alcohol	41.0
N-Propyl Acetate	10.2
Santicizer 141	1.0
Kemamide E	1.0
Polymist A-12	1.0

Nitrocellulose Modification-HMS Ink

	Wt%
Day-Glo HMS Soluble Toner	35.0
1/4 Sec. SS Nitrocellulose (70% dry basis in Ethanol)	10.8
Anhydrous Ethyl Alcohol	38.4
N-Propyl Acetate	12.8
Santicizer 141	1.0
Kemamide E	1.0
Polymist A-12	1.0

Water-Based Flexo Ink

	Wt%
Day-Glo EPX Dispersion	75.4
(adjust pH to 7.5 to 8.0 before adding other ingredients)	
Rheolate 101 (Thickener)	0.4
(disperse ingredients, at high speed, before adding ammonia)	
Ammonia	0.8
G-Cryl 250 (Acrylic Binder)	10.0
Lucidene 602 (Acrylic Binder)	10.0
Tamol 731 (Surfactant)	1.4
Liquitron 251 (Wax Dispersion)	2.0
Defoamer, Biocide	As Needed

SOURCE: Day-Glo Color Corp.: Suggested Formulations

Flexographic Inks

The following formulations are recommended as a starting point for flexographic inks to be used in surface printing of treated polyethylene. Modification may be necessary for individual applications.

Propionate Modification-HM Ink

	Wt%
Day-Glo HM Soluble Toner	35.0
Cellulose Acetate Propionate (CAP-504-0.2)	8.0
Krumbhaar Resin K-1717-B	2.0
Anhydrous Ethyl Alcohol	41.6
N-Propyl Acetate	10.4
Santicizer 141	1.0
Kemamide E	1.0
Polymist A-12	1.0

Propionate Modification-HMS Ink

	Wt%
Day-Glo HMS Soluble Toner	35.0
Cellulose Acetate Propionate (CAP-504-0.2)	8.0
Krumbhaar Resin K-1717-B	2.0
Anhydrous Ethyl Alcohol	39.0
N-Propyl Acetate	13.0
Santicizer 141	1.0
Kemamide E	1.0
Polymist A-12	1.0

T-Type Gravure Ink

	Wt%
Day-Glo T- or GT-Pigment	40.0
Parlon S-20	11.0
Toluene	42.0
Paraplex G-60	5.0
S-395-N5 Wax	2.0

SOURCE: Day-Glo Color Corp.: Suggested Formulations

Flexo Ink
Propionate Modification-HM Ink

	Wt%
Day-Glo HM Soluble Toner	35.0
Cellulose Acetate Propionate (CAP-504-0.2)	8.0
Krumbhaar Resin K-1717-B	2.0
Anhydrous Ethyl Alcohol	41.6
N-Propyl Acetate	10.4
Santicizer 141	1.0
Kemamide E	1.0
Polymist A-12	1.0

Propionate Modification-HMS Ink

	Wt%
Day-Glo HMS Soluble Toner	35.0
Cellulose Acetate Propionate (CAP-504-0.2)	8.0
Krumbhaar Resin K-1717-B	2.0
Anhydrous Ethyl Alcohol	39.0
N-Propyl Acetate	13.0
Santicizer 141	1.0
Kemamide E	1.0
Polymist A-12	1.0

Nitrocellulose Modification-HM Ink

	Wt%
Day-Glo HM Soluble Toner	35.0
1/4 Sec. SS Nitrocellulose (70% dry basis in Ethanol)	10.8
Anhydrous Ethyl Alcohol	41.0
N-Propyl Acetate	10.2
Santicizer 141	1.0
Kemamide E	1.0
Polymist A-12	1.0

Nitrocellulose Modification-HMS Ink

	Wt%
Day-Glo HMS Soluble Toner	35.0
1/4 Sec. SS Nitrocellulose (70% dry basis in Ethanol)	10.8
Anhydrous Ethyl Alcohol	38.4
N-Propyl Acetate	12.8
Santicizer 141	1.0
Kemamide E	1.0
Polymist A-12	1.0

SOURCE: Day-Glo Color Corp.: Suggested Formulations

Formulation Guidelines for CAP in Liquid Inks

Table 3 shows the resins and plastisers evaluated in the formulation. The acrylic resin is added to improve adhesion of the ink to the substrate and adhesion of the PVDC overprint to the ink. The CAP provides fast drying, heat resistance, and low odour; the plasticiser, apart from improving flexibility, is added to reduce film curl, which takes place when the solvent evaporates.

The tendency to curl occurs when the ink film shrinks during solvent evaporation. Although adding plasticiser prevents it, the ink formulator must be careful with the amount of plasticiser used. Too much plasticiser will result in much-diminished heat resistance and will also reduce adhesion in certain cases.

Table 3
Modifying Resins and Plasticisers for CAP-Based Inks on Inks on PVDC-Coated Films

Resins:	Plasticisers
Paraloid B72 acrylic resin	DOP
Plexigum N80 acrylic resin	DIOP
Plexigum N742 acrylic resin	DBP
Neocryl B814 acrylic resin	DMEP
Neocryl B722 acrylic resin	

The top ink in Table 2 contains CAP-504-0.2 in a 4:1 alcohol: ester solvent system, while the base ink uses CAP-482-0.5 in a 1:1 alcohol-ester system. The reason for this was twofold:

*CAP-504-0.2 can produce improved pigment dispersions over CAP-482-0.5, so its use with expensive coloured pigments is sometimes an advantage.

*The less aggressive solvent blend of the top ink will result in less redissolving of the base ink, which may lead to improved gloss.

However, it is perfectly satisfactory to print with either the same CAP grade in both top and base inks or to use them the opposite wav round to that shown in the Table 2.

SOURCE: Eastman Chemical Co.: Suggested Formulations

Gloss Tag & Label Ink-Blue

	Wt%
Dispersion A	35.0
Joncryl 89	31.5
Joncryl 77	25.0
Jonwax 22	5.0
Antifoam: Bubble Breaker 748	0.5
Water	3.0

pH: 8.4
Initial Viscosity*, secs: 26
One Week Viscosity*:
Before agitation: 33
After agitation: 28

Gloss Tag & Label Ink-Yellow

	Wt%
Dispersion B	35.0
Joncryl 89	30.8
Joncryl 77	24.7
Jonwax 22	5.0
Antifoam: Bubble Breaker 748	0.5
Water	4.0

pH: 8.4
Initial Viscosity*, secs: 25
One Week Viscosity*:
Before agitation: 32
After agitation: 25

Gloss Tag & Label Ink-Red

	Wt%
Dispersion C	40.0
Joncryl 89	30.3
Jonwax 77	24.2
Jonwax 22	5.0
Antifoam: Bubble Breaker 748	0.5

pH: 8.6
Initial Viscosity*, secs: 23
One Week Viscosity*:
Before agitation: 36
After agitation: 27

*#2 Zahn cup, 25C

SOURCE: SCJohnson Polymer: Joncryl 67 Suggested Formulations

Glossy Black Corrugated Ink

Raw Materials:	Wt%
Place into clean ball mill:	
Carboset GA-1387	65.2
Huber 35	7.5
DM Water	11.6
Monoethanolamine	0.5
Foamburst 338	0.2
Printex 35	15.0
Ball mill for 12 hours - check grind	

Formulation Constants:

Total Solids by Weight, %: 38.0
Total Solids by Volume, %: 28.0
pH: 8.9
Weight per gallon, lbs.: 9.7
Viscosity, #3 Zahn, 25C, sec.: 25
Pigment to Binder Ratio: 1/1

Formulation WI 152

Overprint Varnish for Wet Litho Ink

Raw Materials:	Wt%
Place into clean mixing vessel:	
Carboset XPD-1993	52.5
Add while mixing:	
Carboset XPD-1931	21.8
DOSS OT-75	2.2
Isopropanol	1.7
Benzoflex 50	1.7
Propylene Glycol	2.6
Water	17.5

Formulation Constants:

Total Solids by Weight, %: 36
pH: 8.7
Print Viscosity, Zahn #3, seconds: 25
% Water Added: 20.0

Roll out with 180-line hand proofer over double bump of litho ink applied with quick-peek proofer.

Metallized paper: 72
SBS board: 54

Formulation CL-231

SOURCE: BF Goodrich Specialty Polymers: Carboset Formulations

Gold Metallic for Paper and Board

Components:	Wt%
Eastek 2140 polymer dispersion (31321) (a)	54.2
Joncryl 1535 acrylic emulsion (b)	15.0
N735A nonleafing aluminum paste (c)	20.0
C.I. Yellow 74 pigment dispersion	9.0
Flexonyl Rubine A-F6B (C.I. Red 184) (d)	1.5
Defoamer (e)	0.2
Biocide (f)	0.1

(a) Eastman
(b) S.C. Johnson
(c) U.S. Aluminum
(d) Hoechst-Celanese
(e) Suggested defoamers include DC 65 Additive (Dow Corning); Surfynol DF-75 (Air Products); Dee Fo 97-3 (Ultra Additives); Defoamer RD-1386 (R&A Auxiliaries); and Mazu DF 327C (PPG).
(f) Suggested biocides include Tektamer 38LV (Calgon); and Proxel GXL and Proxel BD (Zeneca).

Formulation Notes:

1. Components are listed in order of addition.
2. Caution: Excessive off-gassing may result if total alcohol level exceeds about 7 wt %. See Eastman Publication SPI-5 for additional information regarding the safe storage and handling of metallic inks.
3. Typical printing viscosity is 25-30 sec/#2 Zahn EZ series cup. If necessary to adjust viscosity, this formulation can be cut with water.

Application: Gravure

Substrates: Paper
SBS board
Beverage board

Features:

* Metallic appearance
* Adhesion to substrate
* Fast running speeds

SOURCE: Eastman Chemical Co.: Eastek Standard No. 90706

Gravure Ink for Gift Wrap

Components:	Wt%
Eastek 2140 polymer dispersion (31321) (a)	39.2
Eastek 1300 polymer dispersion (31281) (a)	15.3
n-Propyl alcohol	4.0
Pigment dispersions (b)	40.8
Surfynol 104 PA surfactant (c)	0.5
Defoamer (d)	0.1
Biocide (e)	0.1

(a) Eastman
(b) Combinations of Eastek polymer millbases and/or other suggested pigment dispersions to achieve desired color matches.
See Eastman Publication SPI-33.
(c) Air Products
(d) Suggested defoamers include DC 65 Additive (Dow Corning); Surfynol DF-75 (Air Products); Dee Fo 97-3 (Ultra Additives); Defoamer RD-1386 (R&A Auxiliaries); Mazu DF 327C (PPG).
(e) Suggested biocides include Tektamer 38LV (Calgon); Proxel GXL and Proxel BD (Zeneca).

Formulation Notes:
1. Components are listed in order of addition.
2. Typical printing viscosity is 20-25 sec/#2 Shell cup. If necessary to adjust viscosity; this formulation may be cut with water or with a 80/20 blend of water/n-propyl alcohol.

Application: Gravure

Substrates:
Gift wrap
Cigarette labels
Other coated papers

Features:
*Fast running speeds
*User-friendly
*No pH adjustments needed

SOURCE: Eastman Chemical Products: Eastek Standard No. 91477

Green Polyethylene Ink

Raw Materials:	Wt%
Place into a clean mixing vessel:	
Carboset GA-1339	55.0
Premix and add with agitation:	
Water	10.3
Isopropyl Alcohol	5.0
Dowanol DB	4.0
BYK 321	0.5
With moderate agitation add:	
FC 129	0.2
CL 159 (see below)	5.0
Flexiverse Green	20.0
Mix for 10 minutes at slow speed.	

Formulation Constants:
Total Solids by Weight, %: 45.8
pH: 8.7
Viscosity, Zahn #2, 25C, seconds: 26
Formulation WI 201

Polyethylene Wax Dispersion

Raw Materials:	Wt%
Place into a clean mixing vessel:	
Carboset GA-1166	65.0
Add slowly while mixing at high speed.	
Water	4.5
Normal Propyl Alcohol	5.0
Foamburst 157	0.5
Shamrock Neptune N-1	25.0
Continue mixing 30 minutes.	

Formulation Constants:
Total Solids by Weight, %: 46.0
Formulation CL 159

SOURCE: BF Goodrich Specialty Chemicals: Carboset Formulations

High Gloss Foil Ink-Blue

	Wt%
Dispersion A	35.0
Joncryl 2660	53.0
Jonwax 26	5.0
Antifoam: Bubble Breaker 748	0.5
Water	6.5

pH: 8.7
Initial Viscosity*, secs.: 25
One Week Viscosity*:
Before agitation: 28
After agitation: 25

High Gloss Foil Ink-Yellow

	Wt%
Dispersion B	35.0
Joncryl 2660	53.0
Jonwax 26	5.0
Antifoam: Bubble Breaker 748	0.5
Water	6.5

pH: 8.7
Initial Viscosity*, secs.: 25
One Week Viscosity*:
Before agitation: 26
After agitation: 25

High Gloss Foil Ink-Red

	Wt%
Dispersion C	40.0
Joncryl 2660	52.0
Jonwax 26	5.0
Antifoam: Bubble Breaker 748	0.5
Water	2.5

pH: 8.8
Initial Viscosity*, secs.: 25
One Week Viscosity*:
Before agitation: 34
After agitation: 27

*#2 Zahn cup, 25C

SOURCE: SCJohnson Polymer: Joncryl 67 Suggested Formulations

High Gloss Wide Web Gravure Ink-Blue

	Wt%
Dispersion A	35.0
Joncryl 89	42.5
Jonwax 26	5.0
Joncryl 58	10.0
Antifoam: Bubble Breaker 748	0.5
Water	7.0

pH: 8.5
Initial Viscosity*, secs.: 27
One Week Viscosity*:
Before agitation: 39
After agitation: 34

High Gloss Wide Web Gravure Ink-Yellow

	Wt%
Dispersion B	35.0
Joncryl 89	42.5
Jonwax 26	5.0
Joncryl 58	10.0
Antifoam: Bubble Breaker 748	0.5
Water	7.0

pH: 8.5
Initial Viscosity*, secs.: 27
One Week Viscosity*:
Before agitation: 38
After agitation: 32

High Gloss Wide Web Gravure Ink-Red

	Wt%
Dispersion C	40.0
Joncryl 89	42.0
Jonwax 26	5.0
Joncryl 58	10.0
Antifoam: Bubble Breaker 748	0.5
Water	2.5

pH: 8.6
Initial Viscosity*, secs.: 27
One Week Viscosity*:
Before agitation: 53
After agitation: 40

*#2 Zahn cup, 25C

SOURCE: SCJohnson Polymer: Joncryl 67 Suggested Formulations

High Quality Corrugated Ink-Blue

	Wt%
Dispersion A	37.0
Joncryl 74	11.0
Joncryl 87	42.0
Jonwax 26	5.0
Antifoam: Bubble Breaker 748	0.5
Water	4.5

pH: 8.4
Initial Viscosity*, secs.: 25
One Week Viscosity*:
Before agitation: 28
After agitation: 27

High Quality Corrugated Ink-Yellow

	Wt%
Dispersion B	37.0
Joncryl 74	11.0
Joncryl 87	42.0
Jonwax 26	5.0
Antifoam: Bubble Breaker 748	0.5
Water	4.5

pH: 8.4
Initial Viscosity*, secs.: 26
One Week Viscosity*:
Before agitation: 36
After agitation: 32

High Quality Corrugated Ink-Red

	Wt%
Dispersion C	43.0
Joncryl 74	11.0
Joncryl 87	39.0
Jonwax 26	5.0
Antifoam: Bubble Breaker 748	0.5
Water	1.5

pH: 8.6
Initial Viscosity*, secs: 25
One Week Viscosity*:
Before agitation: 35
After agitation: 31

*#2 Zahn cup, 25C

SOURCE: SCJohnson Polymer: Joncryl 67 Suggested Formulations

High Solids, Low Tack Sheetfed Offset Ink

	Wt%
Day-Glo Invisible Blue MT-6172-IB	48.0
AVO-797 #7 Isophthalic Alkyd	34.0
CC-5071 100% Solids PTFE/PE Sheetfed Compound	8.0
Day-Glo Velex Tack Reducer TR-052	8.0
6% Cerium Octoate Drier	2.0

Tack=12.6 @ 1200 RPM/90F
10 Min. Tack Rise=3.0 @ 1200 RPM/90F

Web Offset Heatset Ink

	Wt%
Day-Glo Invisible Blue MT-6172-IB	48.0
Ecogel H Heavy Gel, Low Energy Heatset Web-Offset Vehicle	36.0
CC-6273D Deodorized High Solids PTFE/Wax	8.0
Magiesol 47 Oil	6.0
Kodaflex TXIB	2.0

Tack=10.8 @ 1200 RPM/90F
10 Min. Tack Rise=11.0 @ 1200 RPM/90F

Fluorescent Blue Sheetfed Offset Ink

	Wt%
Day-Glo Invisible Blue MT-6172-IB	56.0
70-MM-2230 Green Shade Phthalo Blue Flush	24.0
TTDIA Polymerized Tung Oil	11.0
CC-5071 100% Solids PTFE/PE Sheetfed Compound	8.0
6% Cerium Octoate Drier	0.4
6% Manganese Lin-All PI Drier	0.4
6% Cobalt Lin-All PI Drier	0.2

Tack=13.0 @ 1200 RPM/90F
10 Min. Tack Rise=4.2 @ 1200 RPM/90F

When formulating invisible blue inks, raw materials such as varnishes and wax compounds should be selected for the lightest possible color and for non-yellowing characteristics so the ink film will not yellow or darken. Cerium drier is recommended for the least amount of color change.

SOURCE: Day-Glo Color Corp.: Suggested Formulations

Ink Compatibility Chart

Ink-Phthalo Blue

	Wt%
Phthalo Blue Dispersion: Flexiverse BFD 1121	40.0
Joncryl 2660	45.0
Jonwax 26	10.0
Water	5.0

Stability Data:
- Initial Viscosity, cps: 45
- 24 Hour Viscosity, cps: 45
- One Week Viscosity, cps: 50
- pH: 8.3

Ink-Phthalo Blue

	Wt%
Phthalo Blue Dispersion: Flexiverse BFD 1121	40.0
Joncryl 2660	45.0
Jonwax 26	10.0
Isopropyl Alcohol	5.0

Stability Data:
- Initial Viscosity, cps: 55
- 24 Hour Viscosity, cps: 90
- One Week Viscosity, cps: 105
- pH: 8.3

Ink-Phthalo Blue

	Wt%
Phthalo Blue Dispersion: Flexiverse BFD 1121	40.0
Joncryl 2660	45.0
Jonwax 26	10.0
Normal Propyl Alcohol	5.0

Stability Data:
- Initial Viscosity, cps: 65
- 24 Hour Viscosity, cps: 120
- One Week Viscosity, cps: 130
- pH: 8.3

SOURCE: SCJohnson Polymers: Joncryl 2660 Suggested Formulations

Ink Compatibility Chart (Continued)

Ink-Carbon Black

Materials:	Wt%
Carbon Black Dispersion: CDI Black DS14060	40.0
Joncryl 2660	45.0
Jonwax 26	10.0
Water	5.0

Stability Data:
- Initial Viscosity, cps: 150
- 24 Hour Viscosity, cps: 160
- One Week Viscosity, cps: 155
- pH: 8.3

Ink-Carbon Black

Materials:	Wt%
Carbon Black Dispersion: CDI Black BS14950	40.0
Joncryl 2660	45.0
Jonwax 26	10.0
Isopropyl Alcohol	5.0

Stability Data:
- Initial Viscosity, cps: 190
- 24 Hour Viscosity, cps: 225
- One Week Viscosity, cps: 260
- pH: 8.3

Ink-Carbon Black

Materials:	Wt%
Carbon Black Dispersion: CDI Black BS14950	40.0
Joncryl 2660	45.0
Jonwax 26	10.0
Normal Propyl Alcohol	5.0

Stability Data:
- Initial Viscosity, cps: 190
- 24 Hour Viscosity, cps: 290
- One Week Viscosity, cps: 450
- pH: 8.3

SOURCE: SCJohnson Polymers: Joncryl 2660 Suggested Formulations

Ink Compatibility Chart (Continued)

Ink-Lithol Rubine

Materials:	Wt%
Lithol Rubine Dispersion: Drew Liquiflex RD5200	40.0
Joncryl 2660	45.0
Jonwax 26	10.0
Water	5.0

Stability Data:
- Initial Viscosity, cps: 45
- 24 Hour Viscosity, cps: 45
- One Week Viscosity, cps: 55
- pH: 8.3

Ink-Lithol Rubine

Materials:	Wt%
Lithol Rubine Dispersion: Drew Liquiflex RD5200	40.0
Joncryl 2660	45.0
Jonwax 26	10.0
Isopropyl Alcohol	5.0

Stability Data:
- Initial Viscosity, cps: 100
- 24 Hour Viscosity, cps: 270
- One Week Viscosity, cps: 450
- pH: 8.3

Ink-Lithol Rubine

Materials:	Wt%
Lithol Rubine Dispersion: Drew Liquiflex RD5200	40.0
Joncryl 2660	45.0
Jonwax 26	10.0
Normal Propyl Alcohol	5.0

Stability Data:
- Initial Viscosity, cps: 140
- 24 Hour Viscosity, cps: 360
- One Week Viscosity, cps: 690
- pH: 8.3

SOURCE: SCJohnson Polymer: Joncryl 2660 Suggested Formulations

Ink Formulation-Black

Pigment Dispersion:	Wt%
Joncryl 60	8.56
Pigment: Carbon Black Elftex 5	12.00
Water	19.44
Letdown:	
Water	19.75
Anti-Foam: Bubble Breaker 748	0.25
SCX-660	40.00

Performance Data of Inks:
- pH: 8.3
- Initial Viscosity, seconds, Zahn cup #2: 24
- One Week Viscosity, sec. Zahn cup #2: 24
- Hot Mar Resistance, F: 425-450
- 60 Gloss: 22-36
- Sutherland Dry Rub Resistance, Cycles: 100

Performance Data of OPV Applied on Above Inks:
- Static Slide Angle: 20-33
- Hot Mar Resistance, F: 450-500
- 60 Gloss: 35-46
- Sutherland Dry Rub Resistance, Cycles: 200

Ink Formulation-Yellow

Pigment Dispersion:	Wt%
Joncryl 60	11.68
Pigment: Diarylide Yellow, Sunbrite Yellow 12, #273-3556	14.00
Water	14.32
Letdown:	
Water	19.75
Anti-foam: Bubble Breaker 748	0.25
SCX-660	40.00

Performance Data of Inks:
- pH: 8.3
- Initial Viscosity, seconds, Zahn cup #2: 23
- One Week Viscosity, sec. Zahn cup #2: 24
- Hot Mar Resistance, F: 450-475
- 60 Gloss: 14-22
- Sutherland Dry Rub Resistance, Cycles: 100

Performance Data of OPV Applied on Above Inks:
- Static Slide Angle: 16-30
- Hot Mar Resistance, F: 450-500
- 60 Gloss: 35-42
- Sutherland Dry Rub Resistance, Cycles: 200

SOURCE: SCJohnson Polymer: SCX-660 Suggested Formulations

Ink Formulation-Blue

Pigment Dispersion:	Wt%
Joncryl 60	11.68
Pigment: Sunfast Blue 15:3 (Green shade) #249-1282	14.00
Water	14.32
Letdown:	
Water	19.75
Anti-foam: Bubble Breaker 748	0.25
SCX-660	40.00

Performance Data of Inks:
- pH: 8.3
- Initial Viscosity, seconds, Zahn cup #2: 21
- One Week Viscosity, sec. Zahn cup #2: 23
- Hot Mar Resistance, F: 400-450
- 60 Gloss: 20-30
- Sutherland Dry Rub Resistance, Cycles: 100

Performance Data of OPV Applied on Above Inks:
- Static Slide Angle: 20-32
- Hot Mar Resistance, F: 450-500
- 60 Gloss: 35-42
- Sutherland Dry Rub Resistance, Cycles: 200

Ink Formulation-Red

Pigment Dispersion:	Wt%
Joncryl 60	8.56
Pigment: Calcium Red 2B, Sunbrite Red 48:2, #234-0077	12.00
Water	19.44
Letdown:	
Water	19.75
Anti-foam: Bubble Breaker 748	0.25
SCX-660	40.00

Performance Data of Inks:
- pH: 8.3
- Initial Viscosity, seconds, Zahn cup #2: 20
- One Week Viscosity, sec. Zahn Cup #2: 30
- Hot Mar Resistance, F: 450-475
- 60 Gloss: 21-33
- Sutherland Dry Rub Resistance, Cycles: 100

Performance Data of OPV Applied on Above Inks:
- Static Slide Angle: 22-30
- Hot Mar Resistance, F: 450-500
- 60 Gloss: 35-43
- Sutherland Dry Rub Resistance, Cycles: 200

SOURCE: SCJohnson Polymer: SCX-660 Suggested Formulations

Low Cost Corrugated Ink-Blue

	Wt%
Dispersion A	35.0
Joncryl 647	20.6
Jonwax 22	5.0
Antifoam: Bubble Breaker 748	0.2
Water	39.2

pH: 8.3
Initial Viscosity*, secs.: 23
One Week Viscosity*:
Before agitation: 27
After agitation: 23

Low Cost Corrugated Ink-Yellow

	Wt%
Dispersion B	35.0
Joncryl 647	20.5
Jonwax 22	5.0
Antifoam: Bubble Breaker 748	0.2
Water	39.3

pH: 8.3
Initial Viscosity*, secs.: 23
One Week Viscosity*:
Before agitation: 20
After agitation: 19

Low Cost Corrugated Ink-Red

	Wt%
Dispersion C	40.0
Joncryl 647	20.5
Jonwax 22	5.0
Antifoam: Bubble Breaker 748	0.2
Water	34.3

pH: 8.3
Initial Viscosity*, secs.: 23
One Week Viscosity*:
Before agitation: 30
After agitation: 24

*#2 Zahn cup, 25C

SOURCE: SCJohnson Polymer: Joncryl 67 Suggested Formulations

Low Energy Web-Offset Heatset Ink

A suggested starting formulation for a low energy web-offset heatset ink using Starfire Base color is as follows:

	Wt%
Day-Glo Starfire Base Color	70.0
Low Energy Heatset Varnish	23.0
S394-N1 Polyethylene Wax	2.0
SST-3 PTFE Powder	1.0
Magiesol 47 Ink Oil	4.0

Tack=10.0 @ 1200 RPM, 90F

Sheetfed Ink

	Wt%
Day-Glo Starfire Fresh Color Base	80.0
Matte Vehicle	14.0
S394-N1 Polyethylene Wax	2.0
Magiesol 52 Ink Oil	2.4
6% Cerium Octoate Drier	1.0
Dri-RX Drier	0.6

Tack=14-16 @ 1200 RPM, 90F

Web Offset Ink

	Wt%
Day-Glo Starfire Fresh Color Base	70.0
Low Energy Heatset Varnish	23.0
S394-N1 Polyethylene Wax	2.0
SST-3 PTFE Powder	1.0
Magiesol 47 Ink Oil	4.0

Tack=14-16 @ 1200 RPM, 90F

Web Offset Heatset Ink

	Wt%
Day-Glo Special Thermoset Base	70.0
Ecogel H, Heavy Gel, Low Energy Heatset Web Offset Vehicle	14.0
CC-6273D, Deodorized High Solids PTFE/Wax Heatset Compound	6.0
Magiesol 52 Oil	8.0
Magiesol 47 Oil	2.0

Tack=10.8 @ 1200 RPM, 90F
10 Min. Tack Rise = 10.0 @ 1200 RPM, 90F

SOURCE: Day-Glo Color Corp.: Suggested Formulations

Neutralization Formulas

Joncryl 142 is supplied in the acidic form and must be neutralized and diluted before subsequent use in ink formulations.

Following are typical formulas for neutralizing Joncryl 142:

Joncryl 91

	Wt%
Water	32.9
Ammonium Hydroxide, 28%	3.6
Joncryl 142	63.5

pH: 7.7
Viscosity, cps: 6,500
% Polymer Solids: 25.0%

Joncryl 85

	Wt%
Water	18.0
Ammonium Hydroxide, 28%	1.8
Isopropyl Alcohol (IPA)	5.0
Joncryl 142	75.2

pH: 6.9
Viscosity, cps: 30,000
% Polymer Solids: 30.0%

A

	Wt%
Water	28.4
Sodium Hydroxide, 25%	8.8
Joncryl 142	62.8

pH: 8.5
Viscosity, cps: 2,200
% Polymer Solids: 27.0%

B

	Wt%
Water	34.5
Monoethanolamine (MEA)	3.0
Joncryl 142	62.5

pH: 8.6
Viscosity, cps: 2,500
% Polymer Solids: 25.4%

SOURCE: SCJohnson Polymer: Joncryl 142 Suggested Formulations

News Flexo Vehicle "H"

Physical Properties:
- Aspect: Transparent Solution
- pH: 7.6-8.1
- Non-Volatile: 28-30%
- Viscosity: 200-400 cps
- Acid Value: 175

Multi-Color News Flexo Vehicle Starting Point Formulations

Red Rubine

	Wt%
Vehicle H	28
Pigment (Presscake, 40%)	51
Water	20
Defoamer	1

Yellow Diarylide

	Wt%
Vehicle H	34
Pigment (Presscake, 40%)	47
Water	18
Defoamer	1

Blue Phthalo Blue

	Wt%
Vehicle H	34
Pigment (Presscake, 40%)	47
Water	18
Defoamer	1

Note: The inks should be run at 15-18 seconds, cup Shell #2

News Flexo Vehicle "F"
News Flexo Vehicle for Water-Based Black Inks Starting Point Formulation

	Wt%
Vehicle F	18
Vulkan K	22
Water	59
Defoamer	1

SOURCE: SCJohnson Polymer: News Flexo Vehicle "H"/News Flexo Vehicle "F": Suggested Formulations

Phthalo G/S Corrugated Ink

Raw Materials:	Wt%
Slowly Premix the following:	
Sunfast Blue, 449-5050	24.0
Carboset GA-1387	65.2
Huber 35	7.5
DM Water	2.6
Monoethanolamine	0.5
Foamburst 338	0.2
Place into a clean ball mill for 12 hours - check grind	

Formulation Constants:

- Total Solids by Weight, %: 35.0
- Total Solids by Volume, %: 26.0
- pH: 9.0
- Weight per gallon, lbs.: 9.4
- Viscosity, #3 Zahn, 25C, sec.: 15
- Pigment to Binder Ratio: 1/1.25

Formulation WI 147

Barium Lithol Corrugated Ink

Raw Materials:	Wt%
Slowly Premix the following:	
R49-HP-1210	24.0
Carboset GA-1387	65.2
Huber 35	7.5
DM Water	2.6
Monoethanolamine	0.5
Foamburst 338	0.2
Place into a clean ball mill for 12 hours - check grind	

Formulation Constants:

- Total Solids by Weight, %: 38.0
- Total Solids by Volume, %: 31.0
- pH: 8.3
- Weight per gallon, lbs.: 9.2
- Viscosity, #3 Zahn, 25C, sec.: 30
- Pigment to Binder Ratio: 1/1.25

Formulation WI 148

SOURCE: BF Goodrich Specialty Chemicals: Carboset Formulations

Pre/Post-Metallization Primer for Paper

Components:	Wt%
Eastek 1200 polymer dispersion (31126) (a)	71.8
Eastek 1300 polymer dispersion (31281) (a)	8.0
n-Propyl alcohol	2.0
Deionized water	17.8
Surfynol 104 PA surfactant (b)	0.3
Biocide (c)	0.1

(a) Eastman
(b) Air Products
(c) Suggested biocides includes Tektamer 38LV (Calgon); Proxel GXL and Proxel BD (Zeneca).

Formulation Notes:

1. Components are listed in order of addition.
2. Typical printing viscosity is 17-20 sec/#2 Zahn EZ series cup. If necessary to adjust viscosity of this formulation, it can be cut with an 80/20 blend of water/n-propyl alcohol.

Application: Gravure
Substrate: Coated paper
Features:
*Resists amine-containing label adhesives
*Promotes glossy metallized surfaces
*Excellent adhesion to vacuum-metallized layer
*Provides a receptive surface for post-metallization printing

SOURCE: Eastman Chemical Co.: Eastek Standard No. 91482

Blue Ink

Materials:	Wt%
GA-2136	47.0
Add under moderate agitation	
Flexiverse Blue BFD-1121	43.0
Michem 39235	5.0
Water	5.0

Formulation Constants:
Viscosity, #2 Zahn, initial, seconds: 33
pH: 8.8
Water added to get print viscosity: 4%
Print viscosity, #2 Zahn, seconds: 25

Defoaming:
Due to the low VOC nature of Carboset GA-2136 foaming is occasionally objectionable. Effective defoamers are:
Foamblast 370 (Ross Chemical)
Foamblast 327 (Ross Chemical)
DeeFo PI-35 (Ultra Additives)
Byk 023* (Byk Chemie)
*Contains silicone

SOURCE: BF Goodrich: Formulation WI-258

Red for Polyethylene

Raw Material:	Wt%
Place into a clean mixing vessel:	
Floxiverse, Red RFD-1133	22.0
With moderate agitation, slowly add:	
Carboset GA-1086	60.0
DM Water	5.8
n-Butanol	2.0
n-Propanol	10.0
FC-120	0.2

Mix with moderate agitation for 30 minutes or until thoroughly mixed.

Formulation Constants:
- Total Solids by Weight, %: 40.1
- Total Solids by Volume, %: 37.9
- pH: 8.5
- Viscosity, Brookfield, 25C, cps: 115
- Weight per Gallon, lbs: 8.6

Formulation WI 144B

High Gloss Water Ink for Paper and Paper Board

Raw Materials:	Wt%
Place into a clean mixer:	
Liqua Flex Predispersed Pigment	30.0
Mix and add slowly while mixing:	
Carboset GA-1087	50.0
Premix and add slowly while mixing:	
Water	7.9
Butyl Cellosolve	4.0
BYK 301	0.1
Mix at low speed for 15 minutes. Continue mixing and add slowly:	
Michem Lube 46025	8.0

Mix at medium speed for 15 minutes.
Adjust viscosity with water. If alcohol is needed, add slowly under agitation.
Formulation WI 105

SOURCE: BF Goodrich Specialty Chemicals: Carboset Formulations

Screen Printing Ink
Vehicle for General Purpose Fluorescent Screen Ink

	Wt%
Aromatic Solvent SC-100 or Super High Flash Naphtha	30.0
EHEC High	6.0
Bentone Gel: Bentone 38 7%; SC-100 90%; 95% Methanol 3%; Made on high speed dissolver	4.0
A hard resin such as Amberol M-82 or Pentalyn G	16.0
360 Mineral Spirits	34.0
Abitol 90X (Can be replaced in part by Castung 403-UV)	10.0

Dissolve resins in the aromatic solvent using a dissolver such as a Cowles and blend in other ingredients in order listed. Strain through nylon filter bags.

Above vehicle	60.0
Day-Glo A-Pigment	35.0
Aromatic Solvent SC-100	5.0

Blend pigment into vehicle using high speed dissolver such as a Cowles and add solvent. Strain color through nylon filter bags. Final viscosity should be about 10,000 cps. Dilution to screening viscosity and wash can be done with mineral spirits.

A Type Gravure Ink
(High Speed Mix)

	Wt%
Acryloid NAD-10	30.0
Day-Glo A- or AX-Pigment	50.0
Toluene	5.0
Thin with Heptane	15.0

C Type Gravure Ink

	Wt%
Day-Glo GT- ot T-Pigment	40.0
RS Nitrocellulose 30-35 Sec. (70% in Toluol)	11.0
Hercolyn D	5.0
S-395-N5 Wax	2.0
Isopropyl Acetate	21.0
Toluene	21.0

SOURCE: Day-Glo Color Corp.: Suggested Formulations

Semi-Permanent Marker Ink

	Wt%
Day-Glo EPX Dispersion	50
(adjust pH to 7.5 to 8.0 before adding other ingredients)	
Propylene Glycol	20
Water	20
Acrylic Binder	10

Textile Ink

	Wt%
Water	68
Day-Glo EPX Dispersion	12
(adjust pH to 7.5 to 8.0 before adding other ingredients)	
Rhoplex E-32 (Acrylic Binder)	12
HiFast 06-59282 (Carrier)	8

Gravure Ink-A Type

	Wt%
Acryloid NAD-10	30.0
Day-Glo AX-Pigment	50.0
Toluene	5.0
Thin with Heptane	15.0

Gravure Ink-C Type

	Wt%
Day-Glo GT- or T-Pigment	40.0
RS Nitrocellulose 30-35 Sec. (70% in Toluol)	11.0
Hercolyn D	5.0
S-395-N5 Wax	2.0
Isopropyl Acetate	21.0
Toluene	21.0

Gravure Ink-T Type

	Wt%
Day-Glo T- or GT-Pigment	40.0
Parlon S-20	11.0
Toluene	42.0
Paraplex G-60	5.0
S-395-N5 Wax	2.0

SOURCE: Day-Glo Color Corp.: Suggested Formulations

Sheetfed Offset

1. For maximum full fluorescent effects when printing primary solid colors:

	Wt%
Starfire Base Color	80.0
Day-Glo Matte Fls. Vehicle, Velex MFV-2080	14.0
S394-N1 Polyethylene Wax	2.0
Magiesol 52 Ink Oil	1.0
6% Cerium Octoate Drier	1.0
12% Manganese Drier	1.0
6% Manganese Drier	1.0

Tack=14-16 @ 1200 RPM, 90F

2. For long printing runs on a wide variety of coated and uncoated paper stock, where maximum press stability, water resistance and sharp halftones are of primary concern:

	Wt%
Starfire Base Color	70.0
Day-Glo Gloss Quickset Vehicle, Velex GQV-2030	20.0
S394-N1 Polyethylene Wax	2.0
Magiesol 52 Ink Oil	5.0
6% Cerium Drier	1.0
12% Manganese Drier	1.0
6% Manganese Drier	1.0

Tack=14-16 @ 1200 RPM, 90F

3. An economical rhodamine replacement for strong, clean spot colors and process printing:

	Wt%
Starfire SFB-22 Strong Magenta	30.0
Lithol Rubine Flush, 45-MM-2247	30.0
Day-Glo Gloss Quickset Vehicle, Velex GQV-2030	29.0
S394-N1 Polyethylene Wax	2.0
Magiesol 52 or Magiesol 60 Oil	5.0
6% Cerium Drier	1.0
12% Manganese Drier	1.0
6% Manganese Drier	1.0

Tack=16-17 @ 1200 RPM, 90F

For even stronger magenta bases, ask for samples of Day-Glo special bases, M-6165-22 (Blue Shade) or M-6210-22 (Yellow Shade)

SOURCE: Day-Glo Color Corp.: Suggested Formulations

Silver Metallic for Paper and Board (Flexo)

Components:	Wt%
Eastek 2140 polymer dispersion (31194) (a)	54.7
Joncryl 1535 acrylic emulsion (b)	15.0
N735A nonleafing aluminum paste (c)	16.7
STAPA hydroxyl W-F 89 N.L. aluminum paste (d)	13.3
Defoamer (e)	0.2
Biocide (f)	0.1

(a) Eastman
(b) S.C. Johnson
(c) U.S. Aluminum
(d) Obrion Atlantic
(e) Suggested defoamers include DC 65 Additive (Dow Corning); Surfynol DF-75 (Air Products); DeeFo 97-3 (Ultra Additives): Defoamer RD-1386 (R&A Auxiliaries); Mazu DF 327C (PPG).
(f) Suggested biocides include Tektamer 38LV (Calgon); Proxel GXL and Proxel BD (Zeneca).

Formulation Notes:
1. Components are listed in order of addition.
2. Caution: Excessive off-gassing may result if total alcohol level exceeds about 7 wt%. See Eastman Publication SPI-5 for additional information regarding safe storage and handling of metallic inks.
3. Typical flexo printing viscosity is about 20 sec/#3 Zahn EZ series cup.

Application: Flexography

Substrates:
Paper
Beverage board

Features:
*Metallic appearance
*Adhesion to substrate and over trapped inks
*Fast running speeds

SOURCE: Eastman Chemical Co.: Eastek Standard No. 90790

Silver Metallic for Paper and Board (Gravure)

Components:	Wt%
Eastek 1100 polymer dispersion (31064) (a)	67.7
Joncryl 1535 acrylic emulsion (b)	15.0
N735A nonleafing aluminum paste (c)	17.0
Defoamer (d)	0.2
Biocide (e)	0.1

(a) Eastman
(b) S.C. Johnson
(c) U.S. Aluminum
(d) Suggested defoamers include DC 65 Additive (Dow Corning); Surfynol DF-75 (Air Products); Dee Fo 97-3 (Ultra Additives); Defoamer RD-1366 (R&A Auxiliaries); and Mafu DF 327C (PPG).
(e) Suggested biocides include Tektamer 38LV (Calgon); and Proxel GXL and Proxel BD (Zeneca).

Formulation Notes:
1. Components are listed in order of addition.
2. Caution: Excessive off-gassing may result if total alcohol level exceeds about 7 wt%. See Eastman Publication SPI-5 for additional information regarding the safe storage and handling of metallic inks.
3. Typical printing viscosity is 25-30 sec./#2 Zahn EZ series cup. If necessary to adjust viscosity, the formulation can be cut with water.

Application: Gravure

Substrates:
Paper
SBS board
Beverage board

Features:
*Metallic appearance
*Adhesion to substrate and over trapped inks
*Fast running speeds

SOURCE: Eastman Chemical Co.: Eastek Standard No. 90465

Stabilized CAP Ink with Polyethylene Imine

Polyethylene Imine:

Polyethylene imine acts as an extremely efficient adhesion promoter for CAP-based inks, probably by forming hydrogen bonds between carboxyl groups on the CAP and amino hydrogen atoms in the polythylene imine. An advantage of this adhesion promoter over the previously described titanium derivatives is that it can be used with low- or zero-acid-number acrylic resins similar to those blended with CAP for printing PVDC-coated films.

A second advantage is that during heat sealing there is no odour from breakdown of the adhesion promoter, as can occur with titanium acetylacetonate products. A disadvantage of polyethylene imine is that unstabilized CAP inks have a shelf life of only one to weeks. However, stabilization of the ink is possible using acetoacetyl-o-toluidine (AAOT) in the ratio of 2 parts AAOT to 1 part polyethylene imine. Table 6 gives a typical formulation.

Table 6

White

Ingredient:	Wt%
Pigment	20.0
CAP 504-0.2	6.7
Paraloid B72	5.8
Silberciser C6	2.0
AAOT	1.0
Ethanol	47.2
Ethyl acetate	16.8
Polymin (water-free)	0.5

Colour

Ingredient:	Wt%
Pigment	10.0
CAP 504-0.2	9.0
Paraloid B72	6.8
Silberciser C6	2.0
AAOT	2.0
Ethanol	54.6
Ethyl acetate	14.6
Polymin (water-free)	1.0

Notes:

*These inks are suitable for reverse printing of coextruded polypropylene for subsequent adhesive lamination.

SOURCE: Eastman Chemical Co.: Suggested Formulations

Starfire Base

Signal Green Base color is not offered as part of the Starfire Base color line, since many variations in color strength may be desirable and are easily obtainable by the ink formulator. The Starfire Signal Green Base color may be formulated as follows:

	Wt%
Day-Glo Saturn Yellow SFB-17	86.7
Blue Shade Phthalocyanine Green Flushed Color, Sun Chemical #U64-2357 (or equivalent)	13.3

Many variations of a fluorescent blue are also obtainable. Day-Glo recommended starting base color is as follows:

	Wt%
Day-Glo Invisible Blue Base MT-6172-IB	70.0
Green Shade Phthalocyanine Blue Flushed Color, BASF #70-MM-2230 (or equivalent)	30.0

1. For maximum full fluorescent effects when printing primary solid colors:

	Wt%
Day-Glo Starfire Base Color	80.0
Matte Vehicle	14.0
S394-N1 Polyethylene Wax	2.0
Magiesol 52 Ink Oil	2.0
6% Cerium Octoate Drier	1.0
12% Manganese Drier	1.0

Tack=14-16 @ 1200 RPM, 90%

2. For long printing runs on a wide variety of coated and uncoated paper stock, where maximum press stability, water resistance and sharp halftones are of primary concern:

	Wt%
Day-Glo Starfire Base Color	70.0
Gloss Quickset Vehicle	20.0
S394-N1 Polyethylene Wax	2.0
Magiesol 52 Ink Oil	6.4
6% Cerium Drier	1.0
12% Manganese Drier	1.0

Tack=14-16 @ 1200 RPM, 90F

3. An economical rhodamine replacement for strong, clean spot colors and process printing:

	Wt%
Day-Glo Starfire SFB-22 Strong Magenta	30.0
Lithol Rubine Flush, 45-MM-2247	30.0
Gloss Quickset Vehicle	29.0
S394-N1 Polyethylene Wax	2.0
Magiesol 52 or Magiesol 60 Oil	6.4
6% Cerium Drier	1.0
12% Manganese Drier	1.0

Tack=16-17 @ 1200 RPM, 90F

SOURCE: Day-Glo Color Corp.: Suggested Formulations

Starting Point Formulations

Joncryl 142 as a sole vehicle system for carbon black inks

Once neutralized, Joncryl 142, performs as a sole vehicle system for carbon blacks. It provides pigment dispersion and transfer properties for corrugated and paper inks.

Pigment Dispersion

	Wt%
Carbon Black: Elftex 8	15.0
Joncryl 91	19.2
Water	25.5
Antifoam: Dee Fo PI-35	0.3

Letdown

Joncryl 91	12.0
Jonwax 22	3.0
Water	25.0

pH: 8.0
Viscosity, #2 Zahn:
Initial: 25 secs
48 hours: 31 secs
% Polymer Solids: 7.8%
% Pigment: 15%

Joncryl 142 as a letdown vehicle:
* Joncryl 142 is a cost-effective letdown vehicle for resin- or surfactant-based organic color dispersions.
* Joncryl 67, Joncryl 678, or Joncryl 690 resins can be used to make vehicles for high quality dispersions. Commercial pigment dispersions, such as those listed, can also be used.

Finished Ink-Yellow

	Wt%
AAA Yellow: Heucosperse YS-1330	40.0
Joncryl 91	26.5
Jonwax 22	3.0
Water	30.0
Antifoam: Bubble Breaker 748	0.5

pH: 8.1
Viscosity, #2 Zahn:
Initial: 25 secs
48 hours: 23 secs

SOURCE: SCJohnson Polymer: Joncryl 142 Suggested Formulations

Starting Point Formulations (Continued)
Joncryl 142 as a Letdown Vehicle (Continued)
Finished Ink-Blue

	Wt%
Phthalo Blue: Flexiverse BFD-1121	35.0
Joncryl 91	35.0
Jonwax 22	3.0
Water	26.5
Antifoam: Bubble Breaker 748	0.5

pH: 8.2
Viscosity, #2 Zahn:
Initial: 24 secs
48 hours: 30 secs

Finished Ink-Red

	Wt%
Barium Lithol: Flexiverse RFD-3217	36.0
Joncryl 91	50.5
Jonwax 22	3.0
Water	10.0
Antifoam: Bubble Breaker 748	0.5

pH: 8.0
Viscosity, #2 Zahn:
Initial: 25 secs
48 hours: 31 secs

Finished Ink-Red

	Wt%
Barium Lithol: Flexiverse RFD-3217	35.0
Joncryl 91	40.0
Jonwax 22	3.0
Water	21.5
Antifoam: Bubble Breaker 748	0.5

pH: 8.0
Viscosity, #2 Zahn:
Initial: 24 secs
48 hours: 26 secs

Finished Ink-Black

	Wt%
Carbon Black: Liquiflex BD-1000	35.0
Joncryl 91	37.0
Jonwax 22	3.0
Water	24.5
Antifoam: Bubble Breaker 748	0.5

pH: 8.2
Viscosity, #2 Zahn:
Initial: 24 secs
48 hours: 28 secs

SOURCE: SCJohnson Polymer: Joncryl 142 Suggested Formulations

Suggested Starting Point Formulations for Finished Inks

Film-Red

	Wt%
Dispersion A	46.0
Joncryl 624	48.5
Jonwax 26	5.0
Antifoam: Bubble Breaker 748	0.5

Initial Viscosity*: 23
One Week Viscosity*: 25

Film-Blue

	Wt%
Dispersion B	43.0
Joncryl 624	51.5
Jonwax 26	5.0
Antifoam: Bubble Breaker 748	0.5

Initial Viscosity*: 25
One Week Viscosity*: 26

Film-Yellow

	Wt%
Dispersion C	43.0
Joncryl 624	51.5
Jonwax 26	5.0
Antifoam: Bubble Breaker 748	0.5

Initial Viscosity*: 25
One Week Viscosity*: 26

*Secs., #2 Zahn Cup

SOURCE: SCJohnson Polymer: Joncryl 678 Suggested Formulations

Suggested Starting Point Formulations for Finished Inks

Foil-Red

	Wt%
Dispersion A	46.0
Joncryl 74	48.5
Jonwax 26	5.0
Antifoam: Bubble Breaker 748	0.5

Initial Viscosity*: 24
One Week Viscosity*: 27

Foil-Blue

	Wt%
Dispersion B	43.0
Joncryl 74	51.5
Jonwax 26	5.0
Antifoam: Bubble Breaker 748	0.5

Initial Viscosity*: 26
One Week Viscosity*: 27

Foil-Yellow

	Wt%
Dispersion C	43.0
Joncryl 74	51.5
Jonwax 26	5.0
Antifoam: Bubble Breaker 748	0.5

Initial Viscosity*: 26
One Week Viscosity*: 27

*Secs, #2 Zahn Cup

SOURCE: SCJohnson Polymer: Joncryl 678 Suggested Formulations

Suggested Starting Point Formulations for Finished Inks

Paper & Board-Red

	Wt%
Dispersion A	46.0
Joncryl 87	47.0
Jonwax 26	5.0
Antifoam: Bubble Breaker 748	0.5
Water	1.5

Initial Viscosity*: 25
One Week Viscosity*: 45

Paper & Board-Blue

	Wt%
Dispersion B	43.0
Joncryl 87	50.0
Jonwax 26	5.0
Antifoam: Bubble Breaker 748	0.5
Water	1.5

Initial Viscosity*: 25
One Week Viscosity*: 28

Paper & Board-Yellow

	Wt%
Dispersion C	43.0
Joncryl 87	50.0
Jonwax 26	5.0
Antifoam: Bubble Breaker 748	0.5
Water	1.5

Initial Viscosity*: 26
One Week Viscosity*: 45

*Secs, #2 Zahn Cup

SOURCE: SCJohnson Polymer: Joncryl 678 Suggested Formulations

Suggested Starting Point Formulations for Finished Inks

Tag & Label-Red

	Wt%
Dispersion A	46.0
Joncryl 87	27.0
Joncryl 77	21.5
Jonwax 26	5.0
Antifoam: Bubble Breaker 748	0.5

Initial Viscosity*: 25
One Week Viscosity*: 31

Tag & Label-Blue

	Wt%
Dispersion B	43.0
Joncryl 87	29.0
Joncryl 77	22.5
Jonwax 26	5.0
Antifoam: Bubble Breaker 748	0.5

Initial Viscosity*: 25
One Week Viscosity*: 27

Tag & Label-Yellow

	Wt%
Dispersion C	43.0
Joncryl 87	29.0
Joncryl 77	22.5
Jonwax 26	5.0
Antifoam: Bubble Breaker 748	0.5

Initial Viscosity*: 25
One Week Viscosity*: 31

*Secs, #2 Zahn Cup

SOURCE: SCJohnson Polymer: Joncryl 678 Suggested Formulations

Transparent Gold for Aluminum Foil

Components:	Wt%
Eastek 1200 polymer dispersion (31125) (a)	88.5
Deionized water	4.9
C.I. Yellow 83 pigment dispersion	4.5
Unisperse Red FBN-PI (C.I. Red 2) (b)	0.5
Wax emulsion (c)	0.5
Surfynol 104 PG surfactant (d)	0.9
Defoamer (e)	0.1
Biocide (f)	0.1

(a) Eastman
(b) Ciba-Geigy
(c) Suggested wax emulsions include Flexonic 600 and Flexonic 901 (Lawter); Hydrocer 600 and Hydrocer 901 (Shamrock Technologies); Jonwax 26 (S.C. Johnson); Michem 46940 and Michem 48040 (Michelman).
(d) Air Products
(e) Suggested defoamers include DC 65 Additive (Dow Corning); Surfynol DF-75 (Air Products); Dee Fo 97-3 (Ultra Additives), Defoamer RD-1386 (R&A Auxiliaries); Mazu DF 327C (PPG).
(f) Suggested biocides include Tektamer 38LV (Calgon); Proxel GXL and Proxel BD (Zeneca)

Formulation Notes:
1. Components are listed in order of addition.
2. Typical printing viscosity is 17-20 sec/#2 Zahn EZ series cup. If necessary to adjust viscosity, this formulation can be cut with water or with an 80/20 blend of water/n-propyl alcohol.

Application: Gravure

Substrate: Aluminum foil

Features:
* Brilliant metallic appearance
* Adhesion to foil
* Resistance to water and menthol/alcohol solutions
* Low odor
* Adjustable shade and COF (coefficient of friction)

SOURCE: Eastman Chemical Co.: Eastek Standard No. 91166

Transparent Light Gold for Aluminum Foil

Components:	Wt%
Eastek 1200 polymer dispersion (31126) (a)	81.4
Eastek 1300 polymer dispersion (31281) (a)	9.0
Deionized water	5.6
Unisperse Red FBN-PI (C.I. Red 2) (b)	0.3
C.I. Yellow 83 pigment dispersion	1.6
Surfynol 104 PG surfactant (c)	1.5
Wax emulsion (d)	0.4
Defoamer (optional) (e)	0.1
Biocide (f)	0.1

(a) Eastman
(b) Ciba-Geigy
(c) Air Products
(d) Suggested wax emulsions include Flexonic 600 and Flexonic 901 (Lawter): Hydrocer 600 and Hydrocer 901 (Shamrock Technologies); Jonwax 26 (S.C. Johnson); Michem 46940 and Michem 48040 (Michelman).
(e) Suggested defoamers include DC 65 Additive (Dow Corning); Surfynol DF-75 (Air Products); Dee Fo 97-3 (Ultra Additives); Defoamer RD-1386 (R&A Auxiliaries); Mazu DF 327C (PPG).
(f) Suggested biocides include Tektamer 38LV (Calgon); Proxel GXL and Proxel BD (Zeneca).

Formulation Notes:
1. Components are listed in order of addition.
2. Typical printing viscosity is 17-20 s/#2 Zahn EZ series cup. If necessary to adjust viscosity, the formulation can be cut with water or with an 80/20 blend of water/n-propyl alcohol.

Application: Gravure

Substrates:
Aluminum foil
Metallized paper

Features:
*Resists menthol/alcohol solutions
*Good substrate adhesion
*Low odor
*Controllable COF (coefficient of friction)

SOURCE: Eastman Chemical: Eastek Standard No. 91602

Treated High Slip, Low Density Polyethylene Film Ink-Blue

	Wt%
Dispersion A	35.0
Joncryl 624	55.0
Jonwax 26	5.0
Antifoam: Bubble Breaker 748	0.5
Water	4.5

pH: 8.4
Initial Viscosity*, secs.: 25
One Week Viscosity*:
Before agitation: 26
After agitation: 26

Treated High Slip, Low Density Polyethylene Film Ink-Yellow

	Wt%
Dispersion B	35.0
Joncryl 624	55.0
Jonwax 26	5.0
Antifoam: Bubble Breaker 748	0.5
Water	4.5

pH: 8.4
Initial Viscosity*, secs.: 25
One Week Viscosity*:
Before agitation: 27
After agitation: 25

Treated High Slip, Low Density Polyethylene Film Ink-Red

	Wt%
Dispersion C	40.0
Joncryl 624	54.5
Jonwax 26	5.0
Antifoam: Bubble Breaker 748	0.5

pH: 8.4
Initial Viscosity*, secs.: 23
One Week Viscosity*:
Before agitation: 32
After agitation: 25

*#2 Zahn cup, 25C

SOURCE: SCJohnson Polymers: Joncryl 67 Suggested Formulations

Victoria Blue Ink for Polyethylene

Raw Materials:	Wt%
Place into a clean mixing vessel:	
Flexiverse BFD 3229	28.0
Slowly mix and add:	
Carboset GA-1086	60.0
Water	2.6
n-Propyl Alcohol	3.0
Michem 32535	6.0
Foamburst 338	0.1
BYK 301	0.2
Fluorad FC120	0.1

Mix at medium speed

Formulation Constants:
Total Solids by Weight, %: 45.8
Total Solids by Volume, %: 42.2
pH: 8.1
Weight per Gallon, lbs: 8.9
Viscosity, #2 Zahn, 25C, seconds: 42
Formula WI 34

Phthalo Blue G/S Ink

Raw Materials:	Wt%
Place into a clean mixing vessel:	
Sunsperse RHD 6017	28.0
Slowly mix and add:	
Carboset GA-1086	60.0
Water	8.6
n-Butanol	3.0
BYK-035	0.1
BYK-301	0.2
Fluorad FC-120	0.1

Mix at medium speed for 15 minutes

Formulation Constants:
Total Solids by Weight, %: 44.5
Total Solids by Volume, %: 43.2
pH: 8.6
Weight per Gallon, lbs: 8.9
Viscosity, #2 Zahn, 25C, seconds: 28
Formula WI 127

SOURCE: BF Goodrich Specialty Chemicals: Carboset Formulations

Yellow Corrugated Ink

Raw Materials:	Wt%
Place into clean ball mill:	
Carboset GA-1387	65.2
Huber 35	7.5
DM Water	14.6
Monoethanolamine	0.5
Foamburst 338	0.2
Sunbrite (AAA)-273-3556	12.0
Ball mill for 12 hours - check grind	

Formulation Constants:
- Total Solids by Weight, %: 35.0
- Total Solids by Volume, %: 28.0
- pH: 9.0
- Weight per Gallon, lbs.: 9.3
- Viscosity, #3 Zahn, 25C, sec.: 15
- Pigment to Binder Ratio: 1/1.25

Formulation WI 150

Black Corrugated Ink

Raw Materials:	Wt%
Place into clean ball mill:	
Carboset GA-1387	65.2
Huber 35	7.5
DM Water	11.6
Monoethanolamine	0.5
Foamburst 338	0.2
Raven 450	15.0
Ball mill for 12 hours - check grind	

Formulation Constants:
- Total Solids by Weight, %: 38.0
- Total Solids by Volume, %: 28.0
- pH: 9.0
- Weight per gallon, lbs.: 9.6
- Viscosity, #3 Zahn, 25C, sec.: 25
- Pigment to Binder Ratio: 1/1

Formulation WI 151

SOURCE: BF Goodrich Specialty Polymers: Carboset Formulations

Yellow Ink

Raw Materials:	Wt%
Place into clean ball mill for 12 hours.	
Carboset GA-1364	69.2
DM Water	18.0
Monoethanolamine	0.3
Foamburst 151	0.4
Foamburst 157	0.1
Sunbrite Yellow 12 (AAA)	12.0
Check grind.	

Formulation Constants:
- Total Solids by Weight, %: 30.0
- Total Solids by Volume, %: 24.0
- pH: 8.6
- Weight per Gallon/lbs.: 9.0
- Viscosity, Brookfield, 25C, cps: 63
- Viscosity, #3 Zahn, 25C, sec.: 30

Formulation WI 160

Blue Ink

Raw Materials:	Wt%
Place into clean ball mill for 12 hours.	
Carboset GA-1364	69.2
DM Water	18.0
Monoethanolamine	0.3
Foamburst 151	0.4
Foamburst 157	0.1
Sunbrite Yellow 12 (AAA)	12.0
Check grind.	

Formulation Constants:
- Total Solids by Weight, %: 30.0
- Total Solids by Volume, %: 23.0
- pH: 8.6
- Weight per Gallon/lbs.: 9.1
- Viscosity, Brookfield, 25C. cps: 45
- Viscosity, #3 Zahn, 25C, sec.: 26

Formulation WI 161

SOURCE: BF Goodrich Specialty Polymers: Carboset Formulations

Section II
Overprint Varnishes

General Purpose OPV (Joncryl 1679)

Primary Purpose:

This is a general purpose OPV with adequate gloss and hardness for most general purpose uses. The formulation can be used as a guide, to allow the formulator flexibility in approaching the more exotic requirements of some OPV's.

Formula:	Wt%
Joncryl 678	8.80
Joncryl 77	59.00
Zinc Oxide Solution #1	5.90
Jonwax 26	10.00
Sodium Hydroxide	1.90
Bubble Breaker 748	0.20
Water	14.20

Formula Joncryl 1679

OPV for Wet Litho (CDX-562)

Primary Purpose:

OPV for in-line and off-line coatings on lithographic printed surfaces using epic-type applicators.

Formula:	Wt%
Joncryl 678	7.0
Joncryl 77	40.0
Joncryl 89	20.0
Jonwax 26	10.0
Ammonia	1.5
IPA	4.0
Water	13.8
Wetting Agent	3.0
Preservative	0.2
Anti-foam	0.5

Typical Characteristics:

Total Non-Volatile: 38.5-40.5%
Viscosity: 100-300 cps
pH: 8.5-9.0
Lbs./Gallon: 8.5-9.0

Formula CDX-562

SOURCE: SC Johnson Polymer: Suggested Formulations

Grease-Barrier Overprints

Table 1 gives a formula for an oil-barrier overprint based on CAP-482-0.5. Table 2 shows a comparison of oil penetration resistance of the CAP-482-0.5 overprint with various other overprints.

Table 1
Grease- and Oil-Barrier Overprint Based on CAP-482-0.5

Components:	Wt%	Lb/100 U.S. Gal.
CAP-482-0.5	21.7	161.2
Eastman Triacetin plasticizer	1.8	13.4
Tecsol C solvent (anhydrous)	53.5	397.5
Ethyl acetate (99%)	23.0	170.9

Properties:
Nonvolatile, wt%: 23.5
Viscosity at 24C (75F):
Brookfield, mPa-s (cP): 576
No. 3 Zahn Cup, sec: 74

The overprint shown in Table 1 may be reduced to application viscosity with a 70/30 weight blend of Tecsol C solvent [(anhydrous)/ethyl acetate (99%).] Triacetin plasticizer, while not mandatory in the formulation, may be used to give flexibility to the film.

Gloss of the overprint may be improved by resin modification with materials such as Acryloid B-66 acrylic resin at approximately two weight parts CAP-482-0.5 to one weight part acrylic resin.

Table 2
Oil Penetration Resistance of Various Overprints (a)

Type of Overprint:	Test Period Hours	Test Temp. C(F)	Oil Penetration Resistance
CAP-482-0.5 (GLRC-4)	2	66(151)	Excellent
Nitrocellulose	2	66(151)	Fair
Polyamide	2	66(151)	Poor
Vinyl chloride-acetate copolymer	2	66(151)	Fair

(a)Overprints were applied to clay-coated paperboard and allowed to dry overnight at room temperature. Corn oil containing a red dye was placed on each overprinted surface and allowed to remain in contact for 2 hours at 66C (151F). The test materials were then wiped off the overprinted surfaces with a soft tissue, and the degree of penetration was observed.

SOURCE: Eastman Chemical Co.: Formula GLRC-4

Heat Resistant OPV

Raw Materials:	Wt%
Place into a clean mixing vessel:	
Carboset GA-1604	53.4
Carboset GA-1166	19.4
Premix and and slowly while mixing:	
Water	16.0
Zinplex 15	4.3
Add while mixing:	
BYK 321	1.1
Michem 160	5.4
BYK 045	0.3
Mix for 30 minutes.	

Formulation Constants:

Total Solids by Weight, %: 31
Viscosity, Brookfield, 25C, cps: 47
Viscosity, #2 Zahn, 25C: 25 seconds
pH: 9.3
Gloss: 45
Blocking, Face-to-Face, 20 psi: 420F
Slide Angle*: 31
Coefficient of Friction (COF): 0.6
*Testing on Carolina 18 stock. Coating applied gravure (165 line anilox) head proofer.

Formulation CL-165

Pet Food Bag OPV

Raw Materials:	Wt%
Carboset GA-1604	60.0
Add the following while mixing:	
15% Zinc Ammonium Carbonate	4.0
Isopropyl Alcohol	5.0
Water	27.0
OP-340	1.0
Liqigard 85	2.0
Santicizer 160	1.0

Physical Properties:

Viscosity, Zahn #2, 25C, secs.: 18
pH: 9.54
Dry Block Test: Pass
Wet Block Test: Pass
Slide Angle: 38.0

Formulation CL-230

SOURCE: BF Goodrich Specialty Polymers: Carboset Formulations

Heat Sealable Overprints

Although CAP-482-0.5 has a high softening range [188C (370F) to 210C (410F)], this cellulose ester can be modified to produce an overprint which heat seals under conditions used in the industry.

Table 5 gives a formulation for a heat-seal overprint for paper. This overprint heat-sealed to itself (face-to-face) at 149C (300F) and 2.8 kg/cm 2 (40 psi), and an unprinted surface (face-to-back) at 177C (350F) and 2.8 kg/cm 2 (40 psi) in three seconds on Kromecoat paper. (1)

Table 5

Components:	Wt%	Lb/100 U.S. Gal
CAP-482-0.5	16.0	124.6
Eastman DOP plasticizer	8.0	62.3
SAIB-90 (Sucrose acetate isobutyrate)	17.7	137.9
Tecsol C solvent (anhydrous)	43.3	337.3
Ethyl acetate (99%)	15.0	116.9

Properties:
Viscosity at 24C (75F):
Brookfield, mPa-s (cP): 422
No. 3 Zahn Cup, sec: 52

If necessary, the overprint solution given in Table 5 may be diluted with a 70/30 blend of Tecsol C solvent (anhydrous) and ethyl acetate (99%). Heat-sealing conditions may be reduced by increasing the DOP and the SAIB concentrations, both of which function to soften the film so that a seal can be formed.
(1) Product of Champion International

Repulpable Overprints

CAP-504-0.2 possesses the ability to form high-gloss overprints which are repulpable under conditions acceptable to the industry.

Paper overprinted with thermoplastic CAP-504-0.2 can be repulped under the following conditions:

1. Heat a mixture of 1 wt% overprinted paper in 99 wt% of a 1 wt% aqueous solution of NaOH at 82C (180F) for 30 minutes.
2. Agitate the mixture (after heating) at moderate speed for 30 minutes.

Heat-Resistant Overprints

Overprints with heat and chemical resistance can be formulated by cross-linking CAP-482-0.5 or CAP-504-0.2 with a reactive resin.
SOURCE: Eastman Chemical Co.: Formula GLRC-5

High Gloss OPV (CDX-563)

Primary Purpose:
Highest gloss OPV for flexo/gravure applications on paper, board, and primed or unprimed foil.

Formula:	Wt%
Joncryl 682	18.0
Joncryl 77	37.0
Joncryl 89	10.0
Ammonia	4.0
Jonwax 26	8.0
Water	13.0
IPA	3.0
Butyl Cellosolve	4.0
Zinc Oxide Solution #1	3.0

Typical Characteristics:
Total Non-Volatile: 40.5-42.5%
Viscosity: 200-800 cps
pH: 8.7-9.3
Lbs./Gallon: 8.7

Formulation CDX-563

Hot Mar Resistant Overprint Varnish

	Wt%
SCX-660	80.00
Anti-Foam: Bubble Breaker 748	0.25
Water	19.75

OPV Performance Data*
pH: 8.3
Viscosity, seconds, Zahn cup #2: 30
Hot Mar Resistance, F: 450
60 Gloss: 48
Static Slide Angle: 24
Sutherland Dry Rub Resistance, Cycles**: 200

Additionally, hot mar resistance can be increased to approximately 500F when SCX-660 is utilized as the ink letdown vehicle and then overprinted with the above varnish formulation.

*Representative performance results can be obtained in the laboratory by simulation of the corrugation heat after application. Performance properties are substrate and ink formulation dependent. Inks have been applied with 180 Q. Anilox. OPV has been applied with 165 Q. Anilox.
**Test results based on various clay coated liners.

SOURCE: SCJohnson Polymer: Suggested Formulations

High-Gloss Thermoset Overprint Based on CAP-504-0.2

Components:	Wt%	Lb/100 U.S. Gal
Part A:		
CAP-504-0.2	7.5	56.4
Cymel 300 hexamethoxymethyl-melamine resin	15.0	112.8
Krumbhaar K-1979 polyester resin	7.5	56.4
BYK 301 (50%) slip additive	2.0	15.0
Tecsol C solvent (anhydrous)	52.5	394.7
Ethyl acetate (99%)	9.5	71.4
Part B:		
p-toluenesulfonic acid (50% in ethyl alcohol)	6.0	45.1

Add Part B to a solution of Part A. This overprint is stable for approximately one month at room temperature and at application viscosity.

Properties:

Viscosity at 24C (75F):

Brookfield, mPa-s (cP): 32

No. 3 Zahn Cup, sec: 20

Formula FCRC-6

Typical Overprint Formulation

Components:	Wt%
Part A:	
CAP-504-0.2	7.8
Resimene U-980	15.5
BYK-301 additive (50%)	1.0
Solvent blend (See below)	74.5
Part B:	
p-Toluenesulfonic acid	1.2

Add Part B to a solution of Part A.

Properties:

Wt% solids: 25.0

Relative properties (5=Best):

Scuff resistance: 5

Gloss: 5

Solvent release: 4

Suggested Solvent Blends:

Blend 1: Ethyl alcohol: 90/Water(1): 10: 100

Blend 2: Isopropyl alcohol: 80/Water(1): 20: 100

Blend 3: n-Propyl alcohol: 70/Water(1):30: 100

Blend 4: Ethyl alcohol: 90/Ethyl acetate: 10: 100

(1) The percentages given are optimum water levels for solubility and film clarity. If desired, water can be changed +/-5% without any detrimental effect.

Formula FCMC-2

SOURCE: Eastman Chemical Co.: Suggested Formulations

High Skid Angle/Low VOC Overprint Varnish

Raw Materials:	Wt%
Place into clean mixing vessel:	
Carboset XPD-1993	63.0
Add while mixing:	
Carboset XPD-1926	25.0
Water	8.0
Surfynol OP-340	2.0
Liquitron 250	2.0

Formulation Constants:

Total Solids by Weight, %: 42.7
pH: 8.7
Weight per gallon, pounds: 8.8
Viscosity, Zahn #2 @ 25C, seconds: 40.0
Drawdown Viscosity, #2 Zahn, seconds: 28.0
% Water Added: 3.0
Gloss on Opacity Chart, 60, #3wwr: 50.0
Gloss*: 43.0
Blocking, 45 psi, Face-to-Back*: Passes 270F
Blocking, 45 psi, Face-to-Face*: Passes 220F
Slide Angle*: 40
Coefficient of Friction (COF)*: 0.816
*Testing performed on Francote 18 stock. Coating applied gravure (150 line anilox) at drawdown viscosity.

Formulation CL-210

Low VOC Overprint Varnish

Raw Materials:	Wt%
Place into clean mixing vessel:	
Carboset XPD-1993	35.0
Add while mixing:	
Carboset XPD-1926	61.0
Liquitron 250	2.0
Surfynol OP-340	2.0

Formulation Constants:

Total Solids by Weight, %: 42.7
pH: 8.9
Weight per Gallon, pounds: 8.8
Viscosity, Brookfield, 25C, cps: 3000
Drawdown Viscosity, #2 Zahn, seconds: 29.0
% Water Added: 21.0
Glass on Opacity Chart, 60, #3wwr: 47.0
Gloss*: 48.0
Blocking, 45 psi, Face-to-Back*: Passes 370F
Blocking, 45 psi, Face-to-Face*: Passes 270F
Slide Angle*: 28
Coefficient of Friction (COF)*: 0.53
*Testing performed on Francote 18 stock. Coating applied gravure (150 line anilox) at drawdown viscosity.

Formulation CL-212

SOURCE: BF Goodrich Specialty Polymers: Carboset Formulations

Low VOC Overprint Varnish

Raw Materials:	Wt%
Place into a clean mixing vessel:	
Carboset GA-2136	63.0
Then add while mixing:	
Carboset GA-1926	25.0
Water	8.0
Surfynol OP-340	2.0
WWX-1041	2.0

Formulation Constants:

Total solids by weight, % (theoretical): 42.8
pH: 8.65
Weight per Gallon, lbs.: 8.9
Viscosity, Brookfield, 25C, cps: 160
Drawdown Viscosity, #2 Zahn, seconds: 25
(% water added): 7
60 Gloss, Carolina 18, #4wwr: 68.5
60 Gloss, Opacity card, #4wwr: 54.0
Blocking, 20 psi, Face-to-Back, F: 235
Blocking, 20 psi, Face-to-Face, F: 215
Slide Angle: 34.5
Coefficient of Friction (COF): 0.68
60 Gloss: 69
Water Resistance(1 Excellent\5 Poor): 1
Alcohol Rub Resistance (1 Excellent\5 Poor): 2

SOURCE: BF Goodrich Specialty Polymers: Formulation CL-234

Offset Ink Train Varnish

Joncryl 74 also lends itself to the formulation of water reducible overprint varnish for offset printing applications. These varnishes can be applied through the existing ink fountain in web offset heatset and sheetfed operations, eliminating the need for a separate roller coater or tower coater. The "open" time of finished products will be directly related to the amine/glycol combination selected allowing for broad formulating parameters. A typical "starting point" formula could include these components:

	Wt%
Joncryl 682	25.0
Propylene Glycol	25.0
Ammonia, 28%	15.0
Joncryl 89	10.0
Jonwax 26	10.0
Joncryl 74	15.0

Initial Viscosity: 690 cps
Viscosity @ 96 hrs: 1600 cps
pH: 10.0
Total Polymer Solids: 38.0%
Gloss, #6 rod on Leneta N2C:
  Black: 90 degrees
  White: 88 degrees
Sutherland, 4 lb x 50: Scratched
Slide Angle, static: 30 degrees
Slide Angle, kinetic: 27 degrees
Dry time*, ambient: 4 hrs. approx.

*Oven assisted dry time supports web offset printing speeds

Procedure:
Dissolve Joncryl 682 in propylene glycol. A laboratory blender and pre-warmed glycol will yield a usable solution in approx imately 10 minutes at high speed. Add ammonia to resin solution. In a separate vessel, add Joncryl 89 to Jonwax 26 and mix to uniformity. Add Joncryl 89/Jonwax 26 blend to resin solution under steady agitation. Finally, add Joncryl 74 to total batch under agitation and mix to uniformity.

SOURCE: SC Johnson Polymer: Joncryl 74 Suggested Formulation

Overprint Lacquer for High-Quality Gift Wrap

Components:	Wt%
Eastek 2140 polymer dispersion (31321) (a)	70.3
Fastek 1300 polymer dispersion (31281) (a)	23.1
n-Propyl alcohol	4.2
Wax emulsion (b)	2.0
Surfynol 104 PA surfactant (c)	0.3
Biocide (d)	0.1

(a) Eastman
(b) Suggested wax emulsions include Flexonic 600 and Flexonic 901 (Lawter); Hydrocer 600 and Hydrocer 901 (Shamrock Technologies); Jonwax 26 (S.C. Johnson); Michem 46940 and Michem 48040 (Michelman).
(c) Air Products
(b) Suggested biocides include Tektamer 38LV (Calgon); Proxel GXL and Proxel BD (Zeneca).

Formulation Notes:
1. Components are listed in order of addition.
2. Typical printing viscosity is 17-20 sec/#2 Zahn EZ series cup. If necessary to adjust viscosity, this formulation can be cut with water or with an 80/20 blend of water/n-propyl alcohol.
3. This formulation is useful as a balanced extender for Eastek 2140 and Eastak 1300 inks.

Application: Gravure

Substrates:
Gift wrap
Coated papers

Features:
*High gloss over acrylic- or solvent-based inks
*Fast running speeds
*No pH adjustments needed

SOURCE: Eastman Chemicals: Eastek Standard No. 91295

Overprint Varnish

Joncryl 67 can also be utilized in overprint varnish formulations to enhance resolubility, transfer, and gloss/leveling properties. The following illustrates a formula that would be appropriate for in-line litho applications.

	Wt%
Jonwax 26	10.0
Joncryl 89	20.0
Joncryl 77	39.0
Joncryl 63	23.5
Isopropanol	4.0
Wetting Agent: Aerosol OT-75	3.0
Antifoam: Bubble Breaker 748	0.5

Performance Data:

Non-Volatile: 39.3%
pH: 8.5
Initial Viscosity, secs (#3 Zahn): 20
60 Gloss on Leneta Form N2A (Black/White): 79/78

Overprint Varnish

Joncryl 56 is an excellent resin solution for overprint varnishes. OPVs using Joncryl 56 can be formulated to improve alkali, water, and wet block resistance.

	% by Weight
Coalesced Joncryl 537*	45.0
Joncryl 56	30.0
N-propyl Alcohol	5.0
Water	10.0
Jonwax 26	10.0

Performance Data:

Viscosity (cps)**: 285
60 Gloss on Leneta 2A-White:
White: 88
Black: 89
Water Resistance (5 minute water spot): Excellent
**Brookfield LVF #2 spindle, 30 rpm. 25C

***Coalesced Joncryl 537:**

Joncryl 537 is an acrylic emulsion that has been specifically designed for its alkali resistance properties. It is a hard polymer that needs to be coalesced (softened) in order to maximize its resistance and film forming properties.

	% by Weight
Joncryl 537	95.0
Glycol Ether: Butyl Cellosolve & Ektasolve EB	5.0

Add the coalesecing solvent to the Joncryl 537 with moderate agitation. Continue agitation until homogeneous.

SOURCE: SCJohnson Polymers: Suggested Formulations

Overprint Varnish over Wet Litho Ink

Raw Materials:	Wt%
Place into a clean mixing vessel:	
Carboset GA-1324	63.5
Add while mixing:	
Carboset GA-1931	26.5
DOSS OT-75	2.6
Isopropanol	2.1
Benzoflex 50	2.1
Propylene Glycol	3.2

Formulation Properties:

pH: 8.6
Weight per Gallon, pounds: 8.7
Print Viscosity, Zahn #3, seconds: 25
% Water Added: 18.0
Gloss, 60C: Rollout with 180-line hand proofer over double bump of litho ink applied with quick-peek proofer.
Metallized Paper: 72
SBS Board: 54

Formulation CL-232

Low VOC Overprint Varnish

Raw Materials:	Wt%
Place into a clean mixing vessel:	
Carboset XPD-1324	63.0
Add while mixing:	
Carboset XPD-1926	25.0
Water	7.9
Surfynol OP-340	2.0
Liquitron 250	2.0
Foamblast 370	0.1

Formulation Constants:

Drawdown Viscosity, #2, seconds: 25
(% water added): (6%)
Gloss on Opacity Chart, 60, #4wwr: 51.3
Gloss on Opacity Chart, 60: 48.9
200 line anilox rollout
Gloss on Carolina 18, 60, #4wwr: 63.1
Slide Angle: 22
Coefficient of Friction (COF): 0.40

Testing performed on Carolina 18 stock. Coating applied #4 WWR a drawdown viscosity.

SOURCE: BF Goodrich Specialty Chemicals: Carboset Formulations

Overprint Varnishes

The following are starting point formulations for high gloss, high solids overprint varnishes made from Joncryl 682 commercially available solutions. The ingredients are listed in the appropriate order of addition.

High Gloss

	Wt%
Jonwax 26	7.7
Water	11.7
Joncryl 89	9.6
Joncryl 77	35.5
Joncryl 52	28.8
Glycol Ether: Butyl Cellosolve & Ektasolve EB	3.8
Zinc Oxide Solution #1	2.9

OPV Performance Data:
- Solids: 40.2%
- Viscosity (cps)*: 230
- Viscosity, #3 Zahn (secs): 16
- 60 Gloss on Leneta N2A:
  - Black: 90
  - White: 87
- pH: 9.1
- Theoretical VOC (by weight): 7.2%
- *Brookfield LVF #2 spindle, 30 rpm, 25C

Low VOC

	Wt%
Jonwax 26	8.2
Water	6.6
Joncryl 89	10.3
Joncryl 77	37.9
Joncryl 50	37.0

OPV Performance Data:
- Solids: 42.9%
- Viscosity (cps)*: 230
- Viscosity, #3 Zahn (secs): 16
- 60 Gloss on Leneta N2A:
  - Black: 81
  - White: 77
- pH: 8.4
- Theoretical VOC (by weight): 0.7%
- *Brookfield LVF #2 spindle, 30 rpm, 25C

SOURCE: SCJohnson Polymer: Joncryl 682 Suggested Formulations

Overprint Varnishes

Solutions of Joncryl 678 are essential components of overprint varnishes that require excellent resolubility, gloss, and hold-out. "Water resistant" refers to the water shedding properties required to manufacture products like fresh produce boxes and paper plates. The ingredients are listed in the recommended order of addition.

Water Resistant

	Wt%
Jonwax 120	15.0
Water	0.9
Antifoam: Bubble Breaker 748	0.1
Joncryl 77	32.5
Joncryl 74	32.5
Joncryl 61	14.0
Zinc Oxide #1	1.5
Silicone Emulsion: Aerosol OT-75	3.5

Non-Volatile: 41.0%
Initial Viscosity, secs (#3 Zahn): 15
Viscosity, 1 week, secs (#3 Zahn): 21
pH: 8.5
Gloss, 60: 62
Sutherland Rub (1=best; 10=worst):
  Face-to-Face: 8
  Face-to-Back: 8
Static Slide Angle:
  Face-to-Face: 30
  Face-to-Back: 29
Kinetic Slide Angle:
  Face-to-Face: 12
  Face-to-Back: 12
Water Shedding: Excellent

SOURCE: SCJohnson Polymer: Joncryl 678 Suggested Formulation

Overprint Varnishes

Solutions of Joncryl 678 are essential components of overprint varnishes that require excellent resolubility, gloss, and hold-out. "In-line litho" is related to the ability of the varnish to level and dry over freshly applied oil based inks. The ingredients are listed in the recommended order of addition.

In-Line Litho

	Wt%
Jonwax 26	10.0
Water	3.5
Joncryl 89	20.0
Antifoam: Bubble Breaker 748	0.5
Joncryl 77	40.0
Joncryl 61	20.0
Wetting/Leveling agent: LE-410	3.0
Isopropanol	3.0

Non-Volatile: 38.5%
Initial Viscosity, secs (#3 Zahn): 16
Viscosity, 1 week, secs (#3 Zahn): 18
pH: 8.4
Gloss, 60: 69
Sutherland Rub (1=best; 10=worst):
  Face-to-Face: 5
  Face-to-Back: 5
Static Slide Angle:
  Face-to-Face: 27
  Face-to-Back: 25
Kinetic Slide Angle:
  Face-to-Face: 16
  Face-to-Back: 17

SOURCE: SCJohnson Polymer: Joncryl 678 Suggested Formulation

Overprint Varnishes

Solutions of Joncryl 678 are essential components of overprint varnishes that require excellent resolubility, gloss, and hold-out. "General purpose" would provide the kind of film integrity necessary for folding cartons, displays, and similar applications. The ingredients are listed in the recommended order of addition.

General Purpose

	Wt%
Jonwax 26	9.7
Water	2.5
Antifoam: Bubble Breaker 748	0.2
Joncryl 77	57.0
Joncryl 60	25.0
Zinc Oxide #1	5.6

Non-Volatile: 38.0%
Initial Viscosity, secs (#3 Zahn): 16
Viscosity, 1 week, secs (#3 Zahn): 16
pH: 9.5
Gloss, 60: 58
Sutherland Rub (1=best, 10=worst):
- Face-to-Face: 7
- Face-to-Back: 7

Static Slide Angle:
- Face-to-Face: 30
- Face-to-Back: 24

Kinetic Slide Angle:
- Face-to-Face: 16
- Face-to-Back: 9

SOURCE: SCJohnson Polymer: Joncryl 678 Suggested Formulations

Overprint Varnishes

Joncryl 74 is an essential component of overprint varnishes that require gloss, holdout, and water and grease resistance. The suggested starting point formulations listed adddess a few specialty applications. "Water resistant" refers to the water shedding properties required to manufacture products like fresh produce boxes and paper plates. "General purpose" would provide the kind of film integrity necessary for folding cartons, displays, and similar applications. The ingredients are listed in the recommended order of addition.

Water Resistant

	Wt%
Jonwax 120	15.0
Water	0.9
Antifoam: Bubble Breaker 748	0.1
Joncryl 74	32.5
Joncryl 77	32.5
Joncryl 61	14.0
Zinc Oxide Solution #1	1.5
Silicone Emulsion: LE-410	3.5

Non-Volatile: 41.0%
Initial Viscosity, secs (#3 Zahn): 13
Viscosity, 1 week, secs (#3 Zahn): 38
pH: 8.8
Gloss*, 60: 56
Sutherland Rub** (1=best; 10=worst):
- Face-to-Face, 500 cycles: 3
- Face-to-Back, 500 cycles: 7

Static Slide Angle*:
- Face-to-Face: 28
- Face-to-Back: 27

Kinetic Slide Angle*:
- Face-to-Face: 10
- Face-to-Back: 11

Water Shedding: Excellent

*Test Substrate=8 pt. Warrenflo Paper (Paperboard)
**Test Substrate=Preprinted litho poster

(Continued)

SOURCE: SC Johnson Polymer: Joncryl 74 Suggested Formulations

Overprint Varnishes (Continued)

General Purpose

	Wt%
Jonwax 26	9.7
Water	2.5
Antifoam: Bubble Breaker 748	0.2
Joncryl 74	57.0
Joncryl 60	25.0
Zinc Oxide Solution #1	5.6

Non-Volatile: 40.0%
Initial Viscosity, secs (#3 Zahn): 16
Viscosity, 1 week, secs (#3 Zahn): 16
pH: 9.2
Gloss*, 60: 49
Sutherland Rub** (1=best; 10=worst):
  Face-to-Face, 500 cycles: 4
  Face-to-Back, 500 cycles: 4
Static Slide Angle*:
  Face-to-Face: 31
  Face-to-Back: 28
Kinetic Slide Angle*:
  Face-to-Face: 17
  Face-to-Back: 16
Water Shedding: NA

*Test Substrate=8 pt. Warrenflo Paper (Paperboard)
**Test Substrate=Preprinted litho poster

SOURCE: SC Johnson Polymer: Joncryl 74 Suggested Formulations

Water & Grease Resistant OPV (Joncryl 1938)

Primary Purpose:

Joncryl 1938 is a water and grease resistant coating for paper plates, food wraps, folding cartons, and furniture shippers. It has good frozen food release, excellent low temperature block resistance, heat seal resistance, and excellent water shedding. It cannot be glued with conventional adhesives.

Formula:	Wt%
Joncryl 77	32.5
Joncryl 74	32.5
Joncryl 61	14.0
Zinc Oxide Solution #1	1.5
Jonwax 120	15.0
Silicone Emulsion*	3.5
Dowicil 75	0.2
Bubble Breaker 748	0.7
Water	0.7

*Union Carbide LE 410 Food Grade Emulsion (avoid silicone fluid).

Components Provide:

Joncryl 77: Film integrity, gloss, hardness for release, and water resistance.

Joncryl 74: Film integrity, gloss, water resistance, oil resistance, chemical resistance, and adhesion.

Joncryl 61: Resolubility, transfer, leveling, and holdout/gloss.

Zinc Oxide Solution #1: Crosslinking for improved chemical resistance, hardness, block resistance, and heat resistance.

Jonwax 120: Paraffin content provides excellent water shedding and release properties; polyethylene promotes gloss and mar resistance.

Silicone: Contributes to all resistance characteristics and provides excellent release properties.

SOURCE: SC Johnson Polymer: Formula Joncryl 1938

Section III
Coatings

Alcohol-Resistant Slipcoat for Aluminum Foil

Components:	Wt.%
Eastek 1200 polymer dispersion (31125) (a)	87.9
Deionized water	10.0
Surfynol 104 PG surfactent (b)	1.5
Wax emulsion (c)	0.5
Biocide (d)	0.1

(a) Eastman
(b) Air Products
(c) Suggested wax emulsions include Flexonic 600 and Flexonic 901 (Lawter); Hydrocer 600 and Hydrocer 901 (Shamrock Technologies); Jonwax 26 (S.C. Johnson); Michem 46940 and 48040 (Michelman).
(d) Suggested biocides include Tektamer 38LV (Calgon); Proxel GXL and Proxel BD (Zeneca).

Formulation Notes:
1. Components are listed in order of addition.
2. Typically, this formulation is cut 30%-50% with water or with an 80/20 water/n-propyl alcohol blend to achieve the desired coating weight and/or COF. Wax content may be adjusted, if necessary, to produce target COF.

Application: Gravure

Substrate: Aluminum foil

Features:
*Resistance to water and menthol/alcohol solutions
*Adhesion to foil
*Low odor
*Controllable COF (coefficient of friction)

SOURCE: Eastman Chemical Co.: Eastex Standard No. 91299

Overprint Varnish

Ingredients:	Total Parts
Polymer	88 Parts
Butyl Cellosolve	6 Parts
Floutron 200	5 Parts
BYK-346	1 Parts

SOURCE: BF Goodrich Co.: Carboset XPD-2224 Formulation

Carboset GA-33 in Waterborne Inks for Vinyl Substrates

Carboset GA-33 resin is particularly suited for use in waterborne printing inks and top lacquers for plasticized vinyl substrates as found in wallcovering, marine and automotive uses. Waterborne inks and top lacquers based on Carboset GA-33 can be formulated to meet forthcoming governmental emission regulations. These coatings can be used on production equipment currently in use. Costly solvent recovery or incineration equipment is not necessary. Inks and coatings based on Carboset GA-33 may also provide a substantial cost savings over solvent borne systems as the price of solvents based on petroleum products continue to rise.

Starting Formulation:

Waterborne inks and coatings made with Carboset GA-33 demonstrate good scrub, abrasion, UV light, and crock resistance, as well as good adhesion and non-blocking characteristics. These coatings can be cured within one minute at 180F when crosslinked with an aziridine crosslinker.

The clear top coat described can be pigmented using commercially available pigment dispersions. It should be noted that not all of the dispersions available to the graphic arts industry will work in this application. It is recommended that the pigment dispersion not contain styrenated acrylics which may lead to long-term adhesion and compatibility problems. Additionally, pigment dispersions or concentrates containing excessive amounts of water sensitive surfactant should be avoided as this may detract from adhesion and resistance properties of the final ink. The suggested starting resin to pigment ratio is approximately 4/1 on a solids basis.

Carboset GA 33
Top Coat for Vinyl Wallcovering

Materials:	Parts per Hundred
Part A:	
Place into a clean mixing vessel.	
Carboset GA-33	80.0
Add while mixing at moderate speed.	
Level 27	6.7
Butyl Cellosolve	8.0
Water	3.7
Part B:	
Add just prior to use.	
XAMA-2*	1.6

Formulation Constants:

Total Solids by Weight, %: 41.7
pH: 7.8
Crosslinker Ratio, resin/aziridine: 4/1
Cure Cycle: 1 minute @ 180F

*When using this ingredient the user should refer to the MSDS

SOURCE: BFGoodrich Specialty Chemicals: Carboset Formulations

Food Release - Bakery Products

Though the normal coating formulations are excellent for most applications, bakery products occasionally stick to the coated surface. For bakery products such as cinnamon rolls, etc., the following formulation shows excellent release.

	Total Solids(%)	Parts (Dry Weight)	Parts (Wet Weight)
Carboset GA-1105	50	60	120
Carboset GA-1103	50	40	80
Cymel 373	50	8	16
Michem Lube 160	25	15	60

To achieve good release, the coating must be crosslinked. The coated surface should reach 300F to achieve crosslinking. If 300F is not possible, a catalyst such as PTSA can be used.

Lidding Films

Lidding films, such as 3M 5850 can be heat sealed to paperboard which has been coated with compounds using Carboset GA-1103 and Carboset GA-1105.

In addition, lidding films can be produced by applying compounds based on the following formulation to an inert film substrate such as polyester.

	Total Solids(%)	Parts(Dry Weight)	Parts(Wet Weight)
Carboset GA-1105	50	70	140
Carboset GA-1103	50	30	60
Propylene Glycol	50	5	10
Sodium Hydroxide	10	to pH 8.5	to pH 8.5
CMC-12M	0.5	0.1	0.2

Economics

Ovenable and non-ovenable coatings are economical.

A coating weight of 6 to 10 lbs./3,000 ft. 2 or 0.3 to 0.5 mils is typically required on both flat board and pre-formed containers.

A coating weight of 24-26 lbs./3,000 ft. 2 ream of PET is used to achieve similar properties.

In addition to the cost of the coating, the equipment needed for applying these specialty emulsions is considerably less expensive than the PET extrusion equipment.

SOURCE: BFGoodrich Specialty Chemicals: Carboset Formulations

Non-Blocking Coatings

Coatings produced using the previously mentioned formulation possess non-blocking characteristics under the following conditions:

* Non-blocking face-to-face at 180F, 2 lbs./in. 2 for 24 hours
* Non-blocking face-to-back at 212F, 2 lbs./in. 2 for 24 hours

To achieve non-blocking characteristics up to 300F, surface waxes, oils and/or crosslinking resins are usually required. A recommended formula is listed below:

	Total Solids(%)	Parts(Dry Weight)	Parts(Wet Weight)
Carboset GA-1105	50	70	140
Carboset GA-1103	50	30	60
Citric Acid	50	1	2
Ammonium Hydroxide	28	to pH 9.5	to pH 9.5
Propylene Glycol	50	7.5	15
Michem Lube 160	25	10	40
Cymel 373	50	10	20
CMC-12M	5	10	200
Ammonium Hydroxide	28	to pH 9.5	to pH 9.5
Brookfield Viscosity		485 cP	485 cP

In order to improve the non-blocking nature of this coating through crosslinking, it must be dried/cured at 300F for 1-2 minutes.

Non-Blocking Moisture Barrier Coating

Foodboard applications occasionally require resistance to moisture vapor as well as oxygen and/or carbon dioxide. A starting formulation for such a coating is listed below:

	Total Solids(%)	Parts(Dry Weight)	Parts(Wet Weight)
Carboset GA-1105	50	60	120
Carboset GA-1103	50	40	80
Michem Lube 316	50	5	10
Michem Lube 743	35	10	29
Propylene Glycol	50	7.5	15
Cymel 373	50	10	20
Citric Acid	100	1	1
MCC-12M	5	0.35	7

Source: BFGoodrich Specialty Chemicals: Carboset Formulations

Repulpability of Paperboard Coated with Carboset GA-1103 and Carboset GA-1105

BFGoodrich has developed two acrylic emulsions, Carboset GA-1103 and Carboset GA-1105, specifically designed for coating ovenable and non-ovenable paperboard in food packaging. Due to environmental concerns, it is desirable to be able to repulp the coated paperboard.

Summary

The Carboset GA-1103/Carboset GA-1105 coated paperboard was repulpable under lab repulping conditions. The coatings were completely separated from the paperboard. The paperboard was completely disintegrated. The Carboset GA-1103/Carboset GA-1105 coatings were not redispersible. They were cut into small pieces in the Waring blender. The coating debris were visible in both the repulped furnish and in the handsheet containing the recycled pulp. In order to use the repulped paperboard, the coating debris needs to be removed from the pulp slurry.

Repulpability of the Coated Paperboard 601-750-637-1

	Wt%
Carboset GA-1103	40
Carboset GA-1105	60

Drying Conditions: 250F 4 minutes
Repulpability: Repulpable
Observation: Pulp contain coating debris & small white particles

601-750-637-2

	Wt%
Carboset GA-1103	40
Carboset GA-1105	60

Drying Conditions: 300F 1 minute
Repulpability: Repulpable
Observation: Pulp contain coating debris & small white particles

SOURCE: BFGoodrich Specialty Chemicals: Carboset Formulations

Repulpability of Paperboard Coated with Carboset GA-1103 and Carboset GA-1105 (Continued)

Repulpability of the Coated Paperboard (Continued)

601-750-637-3

	Wt%
Carboset GA-1103	40
Carboset GA-1105	60
Cymel 373	3

Drying Conditions: 250F 4 minutes
Repulpability: Repulpable
Observation: Pulp contain coating debris & small white particles

601-750-637-4

	Wt%
Carboset GA-1103	40
Carboset GA-1105	60
Cymel 373	3

Drying Conditions: 300F 1 minute
Repulpability: Repulpable
Observation: Pulp contain coating debris & small white particles
Handsheet:
100% Recycyled Pulp: Coating debris on the surface of the handsheet
90% Control Pulp/10% Recycled Pulp: Coating debris on the surface of the handsheet

601-750-637-5

	Wt%
Carboset GA-1103	40
Carboset GA-1105	60

Drying Conditions: 212F 1 minute
Repulpability: Repulpable
Observation: Pulp contain coating debris & small white particles
Handsheet:
100% Recycled Pulp: Coating debris on the surface of the handsheet
90% Control Pulp/10% Recycled Pulp: Coating debris on the surface of the handsheet

Control

	Wt%
Carboset GA-1103	---
Carboset GA-1105	---
Cymel 373	---

Drying Conditions: ---
Repulpability: Repulpable
Observation: Small white particles
Handsheet:
100% Recycled Pulp: Clean
90% Control Pulp/10% Recycled Pulp: ----

SOURCE: BFGoodrich Specialty Chemicals: Carboset Formulations

Vinyl Plastisol Fabric Coating

	Parts per Hundred
High Speed Mix:	
Geon 12	100
DIDP	28
Dioctyl Adipate	12
Drapex 4.4	15
Ferro 1742	6
Pigment A, AX, GT, or T	5-30

Can be converted to an organosol by thinning with volatile diluents such as VM&P Naphtha or Xylene. Fusion Temperature 350F (177C).

The proper stabilizer selection is very important when using fluorescent pigments in PVC. Day-Glo recommends consulting additive manufacturers which frequently offer heavy metal free heat stabilizers specifically designed for use with fluorescent colorants.

Acrylic Urethane Coating

	Wt%
Component A:	
G-Cure 867-Plus 60	16.4
Exempt Solvent Blend	8.9
Day-Glo T-Pigment	35.6

Disperse above ingredients on a high speed mixer and add the following under agitation:

G-Cure 867-Plus 60	17.8
Exempt Solvent Blend	12.3
Additive #11	0.1
Component B Catalyst:	
Desmodur N-75	8.9

Add catalyst to coating before application.

Aerosol Concentrate

	Wt%
Acryloid F-10	40.0
Post 4	1.2
Day-Glo A- or AX-Pigment	35.0
Lactol Spirits	22.0
Toluene	1.8

SOURCE: Day-Glo Color Corp.: Suggested Formulations

Section IV
Miscellaneous

Acrylic Aerosol Concentrate
(High Speed Mix)

	Wt%
Acryloid F-10 (40% in Mineral Thinner/Amsco F) or Elvacite 2044 (40% dissolved in Mineral Spirits)	45.5
Day-Glo A-Pigment	35.0
Lactol Spirits	17.7
Toluene	1.8

Thin with Toluene, or with 1/3 Toluene and 2/3 Lactol Spirits

To can, use 65% of the thinned aerosol concentrate and 35% of a propellant, such as A-70 (propane/isobutane blend)

Alkyd Brushing Paint
(High Speed Mix)

	Wt%
Wallkyd 11-151 (35% in Odorless Mineral Spirits)	44.000
Cyasorb UV 24	0.400
Day-Glo A-Pigment	50.500
12% Cobalt Octoate Drier	0.075
Exkin 2	0.200
Cab-O-Sil	0.200
Mineral Spirits	4.625

Water Dispersion

	Wt%
Water	26.1
2% QP-4400	22.5
Nalco 2341	0.2
Nalco 2335	1.2
Day Glo A- or AX-Pigment	50.0

Add materials in order listed and completely disperse before making next addition. Should be made on a high speed dissolver.

Aerosol Concentrate

	Wt%
Acryloid F-10	44.6
Post 4	1.2
Day-Glo GT- or T-Pigment	34.8
Lactol Spirits	17.6
Toluene	1.8

SOURCE: Day-Glo Color Corp.: Suggested Formulations

Carboset GA-1162 Resin Cut
CL 119

Raw Materials:	Wt%
Place into a clean mixing vessel:	
Water	52.0
Ammonia, 28%	8.0
Isopropyl Alcohol	5.0
Diethylene Glycol Ethyl Ether	3.0
Mix and add slowly while mixing:	
Carboset GA-1162	32.0
Mix for 1 hour or until completely dissolved.	

Formulation Constants:
- Total Solids by Weight, %: 34.5
- Total Solids by Volume, %: 31
- pH: 8.8
- Viscosity, Brookfield, 25C, cps: 140
- Weight per Gallon, lbs: 8.8
- Specific Gravity: 1.05

Carboset GA-1162
Polyethylene Wax Dispersion
CL 135

Material:	Wt%
Place into a clean Cowles mixer:	
CL 119 (See above Formulation)	60.0
Water	15.0
Mix at moderate speed and slowly add:	
S-232	25.0
Mix at high speed for 30 minutes	

Formulation Constants:
- Total Solids by Weight, %: 46.0
- Active Wax by Weight, %: 25.0

SOURCE: BF Goodrich Specialty Chemicals: Carboset Formulations

Carboset GA-1250 Resin Cut

Raw Materials:	Wt%
Place into a clean mixing vessel:	
Water	51.0
Ammonia, 28%	7.5
Isopropyl Alcohol	5.0
Ethylene Glycol	1.5
Diethylene Glycol Ethyl Ether	2.5
Mix and add slowly while mixing.	
Carboset GA-1250	32.5

Formulation Constants:
Total Solids by Weight, %: 36.0
Total Solids by Volume, %: 32.0
pH: 7.8
Brookfield Viscosity, 25C, cps: 2,140
Weight per Gallon, lbs.: 8.8
Specific Gravity @ 25C: 1.06
Formulation CL 121

Carboset GA-1250 Organic Pigment Grinds

Raw Materials:	Wt%
Place into a clean ball mill.	
Carboset GA-1250 Resin Cut (CL 121)	40.0
Water	29.5
BYK 035	0.5
Adjust pH to 8.5 - then add:	
Organic Pigment	30.0
Ball mill 12 hours - check grind.	
Grind should be less than 0.5 mil.	

Formulation Constants:
Total Solids by Weight, %: 44.1
Pigment to Binder Ratio: 2.14/1.00
Formulation WI 130

SOURCE: BF Goodrich Specialty Chemicals: Carboset Formulations

Extender for Bronze Powders/Pastes

Components:	Wt%
Eastek 2140 polymer dispersion (31321) (a)	67.0
Eastek 1300 polymer dispersion (31281) (b)	15.3
n-Propyl alcohol	7.0
Deionized water	9.2
Surfynol 104E surfactant (b)	0.7
Cobratec 99 benzotriazole (c)	0.3
Defoamer (d)	0.2
Associative thickener (optional) (e)	0.2
Biocide (f)	0.1

(a) Eastman
(b) Air Products
(c) PMC Specialties
(d) Suggested defoamers include DC 65 Additive (Dow Corning); Surfynol DF-75 (Air Products); Dee Fo 97-3 (Ultra Additives); Defoamer RD-1386 (R&A Auxiliaries); Mazu DF 327C (PPG).
(e) Suggested associative thickeners include Acrysol RM-825 (Rohm and Haas); DSX-1514 (Henkel).
(f) Suggested biocides include Tektamer 38LV (Calgon); and Proxel GXL and Proxel BD (Zeneca).

Formulation Notes:
1. Components are listed in order of addition.
2. Typically, 25%-35% bronze powder or paste is stirred into this extender to produce an ink. Ink viscosity will depend on bronze level.

Applications: Gravure
Stir-in bronze inks

Substrates: Gift wrap
Other coated papers

Features:
*Provides good "metallic" appearance
*Fast running speeds
*Easy stir-in of bronze powders/pastes, initially and after storage

SOURCE: Eastman Chemical Products: Eastek Standard No. 91430

High Hiding Acrylic Emulsion for Inks on Natural Kraft

SCX-631 is a controlled particle size emulsion designed to hide the brown background of natural kraft substrates. The use of SCX-631 as a letdown vehicle results in more intense and brighter printed colors. The semi-opaque nature of SCX-631 allows for reduction of the level of pigments as well as extenders.

Key Features & Benefits:
* Increase color intensity
* Cost effective
* High dilutability
* Excellent resolubility and printability
* Broad compatibility
* Very low VOC

Starting Point Formulation: White

Pigment Dispersion:	Wt%
Ti-Pure R-960	60.0
Joncryl 60	5.0
SCX-631	8.0
Water	17.0
Letdown:	
SCX-631	6.0
Jonwax 22	3.0
Water	1.0

Ink Values:
- pH: 7.9
- Viscosity, Initial (#2 Zahn): 23 secs
- Viscosity, One week (#2 Zahn): 23 secs

Starting Point Formulation: Black

Pigment Dispersion:	Wt%
Carbon Black: Elftex 5	15.0
Joncryl 91	19.2
Antifoam: Bubble Breaker 748	0.6
Water	25.2
Letdown:	
SCX-631	5.0
Jonwax 22	3.0
Water	32.0

Ink Values:
- pH: 8.0
- Viscosity, Initial (#2 Zahn): 25 secs
- Viscosity, One week (#2 Zahn): 24 secs

SOURCE: SCJohnson Polymer: SCX-631 Suggested Formulations

High Hiding Acrylic Emulsion for Inks on Natural Kraft (Continued)

Starting Point Formulation-Red

Pigment Dispersion:	Wt%
Calcium Lithol Red: Sunbrite Red 48.2	13.0
Joncryl 60	9.6
Water	17.4
Letdown:	
SCX-631	52.0
Jonwax 22	3.0
Water	5.0

Ink Values:
- pH: 8.4
- Viscosity, Initial (#2 Zahn): 27 secs
- Viscosity, One Week (#2 Zahn): 26 secs

Starting Point Formulation-Blue

Pigment Dispersion:	Wt%
Phthalo Blue: Sunfast Blue 15:3 (Green shade)	13.0
Joncryl 60	9.6
Water	14.5
Letdown:	
SCX-631	55.0
Jonwax 22	3.0
Water	4.9

Ink Values:
- pH: 8.2
- Viscosity, Initial (#2 Zahn): 25 secs
- Viscosity, One Week (#2 Zahn): 27 secs

Starting Point Formulation-Yellow

Pigment Dispersion:	Wt%
Diarylide Yellow: Sunbrite Yellow 12	13.0
Joncryl 60	9.6
Water	14.5
Letdown:	
SCX-631	54.6
Jonwax 22	3.0
Water	5.3

Ink Values:
- pH: 8.3
- Viscosity, Initial (#2 Zahn): 25 secs
- Viscosity, One Week (#2 Zahn): 25 secs

SOURCE: SCJohnson Polymer: SCX-631 Suggested Formulations

Pigment Dispersion for Water-Borne Inks Using Carboset Polymers

Introduction

Pigment dispersion is the process by which pigments are ground to a very small particle size and dispersed into a liquid, for subsequent use in a liquid ink or coating. A pigment dispersion for water-borne printing inks consist primarily of pigment, resin, water. Various additives are also used, including wetting agents and anti-foaming agents.

Pigment Selection

The following pigments account for most of the volume in water-borne inks. All are used on a regular basis with good results.

Barium Lithol	Calcium Lithol
Lithol Rubine	Red Lake C
Barium Red 2B	Phthalo Cyan Blue
Phthalo Cyan Green	Diarylide Yellow AAA
Diarylide Yellow AAOT	Carbon Black
Titanium Dioxide	

Starting Formulations

Formulations for pigment dispersions vary greatly, depending on application requirements. Standard dispersions generally contain about 35% pigment and 9% polymer solids, plus water and additives.

To make high gloss dispersions, polymer content is increased and pigment content is decreased. Due to the price differential in these raw materials, the cost per pound for this dispersion will be slightly lower. However, because less pigment is used, the color value of a high gloss dispersion is also lower, and ink formulators may have to use more pigment dispersion per pound of ink to achieve desired color values.

The following starting point formulations have been used successfully in the BFGoodrich laboratories:

Standard Commercial

	Wt%
Pigment	35.0
Polymer solution (Carboset GA-1166HS)	25.0-28.0
Water	37.0-40.0
Anti-foam agent	0.1- 0.2

High Gloss

	Wt%
Pigment	25.0
Polymer solution (Carboset GA-1116HS)	50.0
Water	25.0
Anti-foam agent	0.1-0.2

SOURCE: BFGoodrich Specialty Chemicals: Carboset Formulations

Pigment Dispersion for Water-Borne Inks Using Carboset Polymers (Continued)

Dispersion Procedure:

Dispersions are generally made in three steps: pre-mix, grinding, finishing. The purpose and procedure involved in each step is as follows:

Step 1. Premix:

In this step, pigment agglomerates are broken down into particles small enough to go through the equipment used in Step 2. To do this, the polymer solution, pigment and water are mixed together under high shear conditions. This can be accomplished using a Cowles or Scholds type mixer. A Kady mill can also be used.

Step 2. Grinding:

Here, the particle size is further reduced for more complete dispersion. This can be accomplished in a grinding mill, using a very fine grinding medium, generally less than 2mm in particle size. Either a horizontal or vertical mill can be used, however BFGoodrich has achieved better results using horizontal mills.

The grinding process should be run "container to container", meaning that it should not be directly recirculated through the mill. This configuration assures that the product receives uniform milling. The throughput rate used will vary depending on the quality of the dispersion being made. For highest gloss dispersions smaller particles are required and the throughput should be kept very low to maximize grinding.

Several companies produce shot mills suitable for pigment grinding. BFGoodrich has seen successful grinding using equipment from Buhler, Netzch, Drais, and Scholds.

Step 3. Finishing:

The finishing step is simply an additional pass through the grinding mill to further disperse the pigment. This pass can be made using the same shot or, to give a glossier appearance, finishing can be through extremely fine shot.

SOURCE: BFGoodrich Specialty Chemicals: Carboset Formulations

Pigment Dispersions

Joncryl 67 provides superior wetting properties for the manufacture of high solids, low viscosity pigment dispersions. Typical starting point formulas for both pigment dispersions and finished inks for a variety of substrates and applications are as follows:

Dispersion A

Formula:	Wt%
Calcium Lithol Red 2B (48.2): SunBrite 48:2	32.5
Joncryl 63	23.9
Water	43.6

pH: 8.5
Viscosity:
Initial, cps: 1,775
90 Days, cps: 1,510
Pigment:Binder Ratio: 4:1

Dispersion B

Formula:	Wt%
Phthalo Blue G.S. (15.3): Sunfast Blue 15:3	37.5
Joncryl 63	31.3
Water	31.2

pH: 8.5
Viscosity:
Initial, cps: 450
90 Days, cps: 440
Pigment:Binder Ratio: 4:1

Dispersion C

Formula:	Wt%
Diarylide Yellow, AAA: SunBrite Yellow 12	37.5
Joncryl 63	31.3
Water	31.2

pH: 8.4
Viscosity:
Initial, cps: 1,240
90 Days, cps: 1,280
Pigment:Binder Ratio: 4:1

Procedure:
1. Preblend batch to uniformity using high speed dispersing equipment.
2. Feed blend into small media mill.
3. Increase speed and disperse to required fineness of grind.
4. Recirculate until all air has been displaced.

SOURCE: SCJohnson Polymer: Joncryl 74 Suggested Formulations

Pigment Dispersions

Joncryl 67 provides superior wetting properties for the manufacture of high solids, low viscosity pigment dispersions. Typical starting point formulas are as follows:

Dispersion A

Formula:	Wt%
Phthalo Blue G.S. (15:3): SunFast Blue 15:3 (G.S.)	37.5
Joncryl 63	31.3
Water	31.2

pH: 8.5
Viscosity:
Initial, cps: 450
90 Days, cps: 440
Pigment:Binder Ratio: 4:1

Dispersion B

Formula:	Wt%
Diarylide Yellow, AAA (12): SunBrite Yellow 12	37.5
Joncryl 63	31.3
Water	31.2

pH: 8.4
Viscosity:
Initial, cps: 1,240
90 Days, cps: 1,280
Pigment:Binder Ratio: 4:1

Dispersion C

Formula:	Wt%
Calcium Lithol Red 2B (48:2): SunBrite Red 48:2	32.5
Joncryl 63	23.9
Water	43.6

pH: 8.5
Viscosity:
Initial, cps: 1,775
90 Days, cps: 1,510
Pigment:Binder Ratio: 4:1

Procedure:
1. Preblend batch to uniformity using high speed dispersing equipment.
2. Feed blend into horizontal, small media mill.
3. Increase speed and disperse to required fineness of grind.
4. Recirculate until all air has been displaced.

SOURCE: SCJohnson Polymer: Joncryl 67 Suggested Formulations

Pigment Dispersions

Solutions of Joncryl 678 provide the rheology and wetting to produce high quality pigment dispersions. The solids level of Joncryl 678 solutions can be adjusted to disperse either dry pigment or "high" and "low" solids presscake.

Dispersion A

	Wt%
Calcium Lithol: SunBrite Red 48:2, #234-0077	33.0
Joncryl 60	24.0
Water	43.3

Viscosity, cps: 1825

Dispersion B

	Wt%
Phthalo Blue: SunFast Blue 15:3 (G.S.), #249-1282	35.0
Joncryl 60	26.0
Water	39.0

Viscosity, cps: 360

Dispersion C

	Wt%
Diarylide Yellow: SunBrite Yellow 12, #273-3556	35.0
Joncryl 60	26.0
Water	39.0

Viscosity, cps: 830

Typical Procedure:

1. Preblend batch to uniformity using high speed dispersing equipment.
2. Feed blend into small media mill.
3. Increase speed and disperse to required fineness of grind.
4. Recirculate until all air has been displaced.

SOURCE: SCJohnson Polymer: Joncryl 678 Suggested Formulations

Primer for Polypropylene and Other Plastics and Metals

Raw Materials:	Wt% 1	Wt% 2
CP 343-1 (25%)	20.0	10.0
Elvax 260 or 40W Resin (1)	----	2.5
Toluene	80.0	87.5

(1) Ethylene vinyl acetate copolymer resins from DuPont.

Stir-In Additive in Coatings for Polypropylene and Other Plastics

Raw Materials:	Wt% 1	Wt% 2
CP 343-3 (50%)	5.0 (1)	----
CP 515-2 (40%)	---	7.5 (1)
Acrylic enamel or lacquer	100.0	100.0

(1) Level of chlorinated polyolefin may be adjusted to give optimum properties.

Primer for Polyethylene

Raw Materials	Wt% 1	Wt% 2
CP 153-2 (25%)	20.0	10.0
Elvax 40W or 260 resin (1)	----	2.5
Toluene	80.0	87.5

(1) Ethylene vinyl acetate copolymer resins from DuPont.

Eastman Chlorinated Polyolefins in Coatings for Selected Plastics:

Features and Benefits:

*Promotes adhesion to polyolefins and other plastics
*Promotes adhesion to aluminum and galvanized steel
*Useful for stir-in or primer formulations
*Useful for inks and coatings
*Available in solution or powder form

SOURCE: Eastman Chemical Co.: Suggested Formulations

Resin Solutions

The following resin solutions are commercially available:

Joncryl 50

	Wt%
Joncryl 682 Resin Flake	50.0
Ammonia, 28%	11.8
Water	38.2

pH: 8.4
Viscosity (cps)*: 5,000

Joncryl 52

	Wt%
Joncryl 682 Resin Flake	60.0
Isopropyl Alcohol	10.0
Ammonia, 28%	14.0
Water	16.0

pH: 8.7
Viscosity (cps)*: 5,000

Joncryl 55

	Wt%
Joncryl 682 Resin Flake	61.0
Ammonia, 28%	14.2
Water	24.8

pH: 8.4
Viscosity (cps)*: 30,000
*Brookfield LVF #3 spindle, 12 rpm, 25C

Procedure for making the resin solutions:

1. Add hot water (130-150F), isopropyl alcohol (if applicable), and ammonia in a covered vessel fitted with high speed dispersing agitation.
2. Start agitation and add Joncryl 682 slow enough to prevent the formation of lumps or balls. Agitate at medium speed for 15 minutes.
3. Cool and check pH. Adjust pH with ammonia to required specification.

Precautions must be taken in order to prevent loss of alcohol or ammonia during the resin cutting procedure. Significant loss of either the alcohol or ammonia may result in undissolved resin in the solution.

Resin cuts can be made at various viscosities by adjusting the solids level, alcohol, or pH.

SOURCE: SCJohnson Polymer: Joncryl 682 Suggested Formulations

Solvent Free Resin Cut

Raw Materials:	Wt%
Place into clean mixing vessel:	
Water	56.0
Ammonia, 27%	9.0
Add slowly while mixing:	
Carboset GA-1161	35.0

Mix at high speed for 45 minutes or until all resin has dissolved

Formulation Constants:
  Total Solids by Weight, %: 35.1
  Viscosity, Brookfield, 25C, cps: 1850
  Weight per Gallon, lbs: 8.9
Formulation CL 101B

Organic Pigment Grinds

Raw Materials:	Wt%
Place into a clean mixing vessel:	
Carboset GA-1161 Solvent Free Resin Cut (Formula CL 101B)	40.0
Water	25.9
BYK 035	0.5
Adjust pH to 8.5-then add:	
Organic Pigment	30.0

Shot mill 12 hours-check grind.
Grind should be less than 0.5 mil.

Formulation Constants:
  Total Solids by Weight, %: 44.2
  Pigment to Binder Ratio: 2.14/1.00
Formulation WI119B

SOURCE: BF Goodrich Specialty Chemicals: Carboset Formulations

Solvent Free Resin Cut

Raw Materials:	Wt%
Place into a clean mixing vessel:	
Water	56.3
Ammonia, 28%	8.7
Add slowly while mixing:	
Carboset GA-1162	35.0

Formulation Constants:
- Total Solids by Weight, %: 35.0
- Total Solids by Volume, %: 33.1
- pH: 8.9
- Viscosity, Brookfield, 25C, cps: 1032
- Weight per Gallon, lbs.: 8.8

Formulation CL 173

Organic Pigment Grinds

Raw Materials:	Wt%
Place into a clean ball mill:	
Carboset GA-1162 Resin Cut (CL 109B)	40.0
Water	29.5
BYK 035	0.5
Adjust pH to 8.5 - then add:	
Organic Pigment	30.0

Ball mill 12 hours - Check grind.
Grind should be less than 0.5 mil.

Formulation Constants:
- Total Solids by Weight, %: 44.1
- Pigment to Binder Ratio: 2.14/1.00

Formulation WI 121B

SOURCE: BF Goodrich Specialty Chemicals: Carboset Formulations

Solutions of Joncryl 678

Joncryl 678 is supplied in solid form and must be neutralized and dissolved to form a varnish before use. Solutions of Joncryl 678 shown below are commercially available as Joncryl 60 and Joncryl 61:

Joncryl 60

	Wt%
Joncryl 678	34.00
Ammonia, 28%	7.50
Water	58.50

Appearance: Clear solution
Non-Volatile: 34.0%
Viscosity: 5500 cps
pH: 8.4
Weight per Gallon: 8.94 lbs.
Theoretical VOC (by Weight): 0.7%

Joncryl 61

	Wt%
Joncryl 678	35.00
Ammonia, 28%	7.40
Propylene Glycol	1.45
Isopropanol	5.65
Water	50.50

Appearance: Clear solution
Non-Volatile: 35.5%
Viscosity: 3250 cps
pH: 8.3
Weight per Gallon: 8.80 lbs.
Theoretical VOC (by Weight): 7.8%

Procedure for making the resin solutions:

1. Add hot water (130-150F), isopropyl alcohol (if applicable), and ammonia in a covered vessel fitted with high speed dispersing agitation.
2. Start agitation and add Joncryl 678 slow enough to prevent the formation of lumps or balls. Agitate at medium speed until dissolved.
3. Cool and check pH. Adjust pH with ammonia to required specification.

Precautions must be taken in order to prevent loss of alcohol or ammonia during the resin cutting procedure. Significant loss of either the alcohol or ammonia may result in undissolved resin in the solution.

SOURCE: SCJohnson Polymer: Joncryl 678 Suggested Formulations

Vinyl Plastisol Molding, Film Foaming & Fabric Coating
(High Speed Mix)

	Parts per Hundred
Geon 121	100
DIDP (Coalescing Solvent)	28
Dioctyl Adipate (Coalescing Solvent)	12
Drapex 4.4 (Stabilizer)	15
Ferro 1742 (Stabilizer)	6
Day-Glo A- or AX-Pigment	5-30

Can be converted to an organosol by thinning with volatile diluents such as VM & P Naphtha or Xylene. Fusion temperature is 350F (177C).

Semi-Rigid Calendered Vinyl Film

Tenneco 225F	100.0
DIDP	15.0 phr
Mark GS	2.5 phr
Drapex 6.8	5.0 phr
Day-Glo GT- or T-Pigment	10.0 phr

Fluorescent Paper Coating

	Wt%
Day-Glo EPX Dispersion	25
(adjust pH to 7.5 to 8.0 before adding other ingredients)	
Calcium Carbonate	15
Thickener	As Needed
Latex Binder	23
Water	37
Defoamer, Biocide	As Needed

Tempera Paint

	Wt%
Day-Glo EPX Dispersion	50
(adjust pH to 7.5 to 8.0 before adding other ingredients)	
Thickener	As Needed
Latex Binder	25
Water	25
Defoamer, Biocide	As Needed

SOURCE: Day-Glo Color Corp.: Suggested Formulations

Section V

Trademarked Raw Materials

RAW MATERIAL	CHEMICAL DESCRIPTION	SOURCE
Abitol 90X	Hydrobietyl alcohol	Hercules
Acryloid B-66	Acrylic resin solution. TSC:51%	Rohm&Haas
Acryloid F-10	Acrylic resin solution. TSC:40%	Rohm&Haas
Acryloid NAD-10	Acrylic dispersion resin.	Rohm&Haas
Acrysol RM-825	Associative thickener	Rohm&Haas
Additive #11		CasChem
Aerosol OT-75	Wetting/leveling agent	AmCyan
Amberol M-82		Rohm&Haas
Aromatic Solvent SC-100	Petroleum solvent. BP: 311F	Exxon
AVO-797 #7	Isophthalic alkyd	BASF
BASF #70-MM-2230	Green shade phthalocyanine blue	BASF
Bentone 38	Rheological additive	Rheox
Benzoflex 50	Plasticizer	Velsicol
Bubble Breaker 748	Antifoam	Witco
Butyl Cellosolve	Glycol ether	UnCarb
BYK 023	Defoamer	BYKChemie
BYK 035	Defoamer	BYKChemie
BYK 045	Defoamer	BYKChemie
BYK 301	Silicone leveling aid	BYKChemie
BYK 321	Flow aid	BYKChemie

RAW MATERIAL	CHEMICAL DESCRIPTION	SOURCE
Cab-O-Sil	Fumed silica	Cabot
Cab-O-Sil M-5	Fumed silica	Cabot
CAP-482-0.5	Cellulose acetate propionate. Vis: 0.40 sec	Eastman
CAP-504-0.2	Cellulose acetate propionate. Vis: 0.15 sec	Eastman
Carbopol Ultrez 10	Carbomer	Goodrich
Carboset GA-33	Acrylic dispersion copolymer. Hard thermoplastic	Goodrich
Carboset GA-1086	Styrene acrylic copolymer emulsion	Goodrich
Carboset GA-1087	Styrene acrylic copolymer emulsion	Goodrich
Carboset GA-1103	Acrylic emulsion. Medium hardness	Goodrich
Carboset GA-1105	Acrylic emulsion. Very hard.	Goodrich
Carboset GA-1161	Styrene-acrylic copolymer. Low MW	Goodrich
Carboset GA-1162	Styrene-acrylic copolymer. Low MW	Goodrich
Carboset GA-1166	Styrene-acrylic polymer solution	Goodrich
Carboset GA-1166HS	Styrene-acrylic polymer solution	Goodrich
Carboset GA-1250	Styrene-acrylic polymer. Low MW	Goodrich
Carboset GA-1324	Acrylic copolymer emulsion	Goodrich
Carboset GA-1339	Styrene-acrylic copolymer emulsion	Goodrich
Carboset GA-1364	Acrylic copolymer dispersion	Goodrich
Carboset GA-1387	Styrene-acrylic copolymer dispersion	Goodrich

RAW MATERIAL	CHEMICAL DESCRIPTION	SOURCE
Carboset GA-1604	Acrylic copolymer emulsion	Goodrich
Carboset GA-1926	Polymer	Goodrich
Carboset GA-1931	Resin solution	Goodrich
Carboset GA-2136	Acrylic copolymer emulsion	Goodrich
Carboset XPD-1324	Polymer	Goodrich
Carboset XPD-1926	Polymer	Goodrich
Carboset XPD-1931	Polymer	Goodrich
Carboset XPD-1993	Styrene-acrylic copolymer emulsion	Goodrich
Castung 403-UV	Dehydrated castor oil	CasChem
CC-5071 100% Solids PTFE/PE Sheetfed Compound		Carroll
CC-6273 Deodorized	High solids PTFE/Wax	Carroll
CDI Black BS14950	Carbon black dispersion	CDIDisp
Cerium Octoate Drier-6%		Shepherd
Cobalt Lin-All PI-Drier-6%		Mooney
Cobratec 99	Benzotriazole	PMCSpec
CP 153-2	Chlorinated paraffin	Eastman
CP 343-1	Chlorinated paraffin	Eastman
CP 343-3	Chlorinated paraffin	Eastman
CP 515-2	Chlorinated paraffin	Eastman
Crodamine OR Wax		Croda
Cymel 300	Hexamethoxymethyl-melamine resin	AmCyan
Cymel 373	Melamine resin	AmCyan
Cyasorb UV24	UV light stabilizer	AmCyan

RAW MATERIAL	CHEMICAL DESCRIPTION	SOURCE
Day-Glo A-Pigment	Thermoplastic fluorescent pigment	Day-Glo
Day-Glo AIT Colorant		Day-Glo
Day-Glo AX Pigment	Thermoplastic fluorescent pigment	Day-Glo
Day-Glo EPX Disp	Formaldehyde free pigment	Day-Glo
Day-Glo GT Pigment	Thermoset resin based	Day-Glo
Day-Glo HM Soluble Toner	Transparent flourescent colorant	Day-Glo
Day-Glo HMS Soluble Toner	Transparent fluorescent colorant	Day-Glo
Day-Glo Invisible Blue Base MT-6172-IB		Day-Glo
Day-Glo Special Base M-6165-22	Magenta base (Blue shade)	Day-Glo
Day-Glo Special Base M-6210-22	Magenta base (Yellow shade)	Day-Glo
Day-Glo Special Thermoset Base		Day-Glo
Day-Glo Starfire Base Color		Day-Glo
Day-Glo Starfire Fresh Color Base		Day-Glo
Day-Glo T-Pigment	Thermoset resin based	Day-Glo
Day-Glo Velex Tack Reducer TR-052		Day-Glo
DC 65 Additive	Defoamer	DowCorn
DeeFo PI-35	Defoamer	UltraAdd
DeeFo 97-3	Defoamer	UltraAdd
Defoamer RD-1386	Defoamer	R&A Auxil
Desmodur N-75	Catalyst	Mobay
DOSS OT-75	Wetting agent	Cytec

RAW MATERIAL	CHEMICAL DESCRIPTION	SOURCE
Dowanol DB	Solvent	Dow
Dowicil 75	Antimicrobial agent	Dow
Drapex 4.4	Stabilizer	Argus
Drapex 6.8	Vinyl plasticizer	Argus
Dri-RX Drier		Mooney
DSX-1514	Associative thickener	Henkel
Dynasilan DAMO	Amino silane derivative	Dynamit
Eastek 1100	Polymer dispersion (31064)	Eastman
Eastek 1200	Polymer dispersion (31125)	Eastman
Eastek 1300	Polymer dispersion	Eastman
Eastek 2140	Polymer dispersion (31321)	Eastman
Eastman DOP	Dioctyl phthalate plasticizer	Eastman
Eastman Triacetin	Plasticizer	Eastman
Ecogel H	Heatset web offset vehicle	Ecovar
EHEC High	Ethyl hydroxyethyl cellulose	Hercules
Ektasolve EB	Glycol ether	Eastman
Elftex 5	Carbon black	Cabot
Elftex 8	Carbon black	Cabot
Elvacite 2044		duPont
Elvax 40W	Ethylene vinyl acetate copolymer	duPont
Elvax 260 Resin	Ethylene vinyl acetate copolymer	duPont
Exkin 2	Anti-skinning agent	Huls

RAW MATERIAL	CHEMICAL DESCRIPTION	SOURCE
FC-120	Wetting agent	3M
FC-129	Wetting agent	3M
Ferro 1742	Stabilizer	Ferro
Flexiverse BFD-1121	Phthalo blue dispersion	Sun
Flexiverse BFD-3229	Pigment dispersion	Sun
Flexiverse Green	Pigment dispersion	Sun
Flexiverse RCD-7108	Barium lithol red (surfactant based)	Sun
Flexiverse RFD-1133	Red pigment dispersion	Sun
Flexiverse RFD-3217	Barium lithol red	Sun
Flexonic 600 & 901	Wax emulsion	Lawter
Flexonyl Rubine A-F6B	C.I. Red 184	Hoechst
Foamblast 327	Defoamer	Ross
Foamblast 370	Defoamer	Ross
Foamburst 151	Defoamer	Ross
Foamburst 157	Defoamer	Ross
Foamburst 338	Defoamer	Ross
G-Cryl 250	Acrylic binder	Henkel
G-Cure 867-Plus 60	Acrylic resin	Henkel
Geon 12	Resin	Goodrich
Geon 121	Resin	Goodrich
Germaben II	Preservative	Sutton

RAW MATERIAL	CHEMICAL DESCRIPTION	SOURCE
Hercolyn D	Hydrogenated rosin ester	Hercules
Heucosperse YS-1330	AAA yellow	Heubach
HiFast 06-59282	Carrier	BASF
Huber 35	Kaolin clay	Huber
Huber 900	Kaolin clay	Huber
Hydrocer 600 & 901	Wax emulsion	Shamrock
Joncryl 50	Acrylic polymer solution	Johnson
Joncryl 52	Acrylic resin solution	Johnson
Joncryl 55	Acrylic resin solution	Johnson
Joncryl 56	Acrylic resin emulsion	Johnson
Joncryl 58	Acrylic resin solution	Johnson
Joncryl 60	Acrylic resin solution	Johnson
Joncryl 61	Resin solution	Johnson
Joncryl 62	Acrylic resin solution	Johnson
Joncryl 63	Acrylic resin solution	Johnson
Joncryl 67	Solid acrylic resin	Johnson
Joncryl 74	Acrylic emulsion	Johnson
Joncryl 77	Acrylic emulsion	Johnson
Joncryl 85	Acrylic colloidal solution	Johnson
Joncryl 87	Styrenated acrylic emulsion	Johnson
Joncryl 89	Styrenated acrylic emulsion	Johnson
Joncryl 91	Acrylic colloidal solution	Johnson
Joncryl 142	Acrylic polymer emulsion	Johnson

RAW MATERIAL	CHEMICAL DESCRIPTION	SOURCE
Joncryl 537	Acrylic polymer emulsion	Johnson
Joncryl 586	Solid acrylic resin	Johnson
Joncryl 624	Acrylic polymer emulsion	Johnson
Joncryl 647	Acrylic colloidal solution	Johnson
Joncryl 678	Solid acrylic resin	Johnson
Joncryl 682	Solid acrylic resin	Johnson
Joncryl 1535	Acrylic emulsion	Johnson
Joncryl 2660	Acrylic polymer emulsion	Johnson
Jonwax 22	Microcrystalline wax emulsion	Johnson
Jonwax 26	Polyethylene wax	Johnson
Jonwax 120	Polyethylene/paraffin wax	Johnson
Kemamide E	Erucamide fatty amide	Humko
Kodaflex TXIB	Plasticizer	Eastman
Krumbhaar Resin K-1717-B		Lawter
Krumbhaar K-1979	Polyester resin	Lawter
Lactol Spirits		
Lancoglidd TD (25%)	Wax	
LE-410	Wetting/leveling agent. Emulsion	UnCarb
Level 27	Flattening agent	
Liqigard 85	Wax	Carroll
Liqua Flex	Predispersed pigment	Drew
Liquiflex BD-1000	Carbon black	Drew

RAW MATERIAL	CHEMICAL DESCRIPTION	SOURCE
Liquaflex RD5200	Lithol rubine dispersion	Drew
Liquaflex YD-3000	Yellow pigment dispersion	Drew
Liquitron 250	Wax	Carroll
Liquitron 251	Wax dispersion	Carroll
Lithol Rubine Flush, 45-MM-2247		BASF
Lucidene 602	Acrylic binder	Morton
Luviskol VA-73W	PVP/VA copolymer (50%)	BASF
Mackamide AME	Acetamide MEA	McIntyre
Magiesol 47 Ink Oil	Aliphatic ink oil	Magie
Magiesol 52 Ink Oil	Aliphatic ink oil	Magie
Magiesol 60 Ink Oil	Aliphatic ink oil	Magie
Manganese Drier-12%	12% manganese	Mooney
Manganese Lin-All PI Drier-	6% manganese	Mooney
Mark GS	Stabilizer	ArgusWitco
Mazu DF 327C	Defoamer	PPG
Merquat Plus 3330	Polyquaternium-39	
Michem Lube 160	Wax	Michelman
Michem Lube 316		Michelman
Michem Lube 743		Michelman
Michem Lube 32535	Wax	Michelman
Michem Lube 46025	Wax	Michelman
Michem 46940	Wax emulsion	Michelman
Michem 48040	Wax emulsion	Michelman
Moisturizing Phytoamine Biocomplex		

RAW MATERIAL	CHEMICAL DESCRIPTION	SOURCE
N735A	Nonleafing aluminum paste	USAluminum
Nalco 2321 Defoamer		Nalco
Nalco 2335 Dispersant		Nalco
Nalco 2341		Nalco
Neocryl B722	Acrylic resin	ICI
Neocryl B814	Acrylic resin	ICI
Neptune N-1	CL159 recipe, 25 parts N-1	Shamrock
OP-340	Surfactant	AirProd
Paraloid B72	Resin	Rohm&Haas
Paraplex G-60	Plasticizer	Rohm&Haas
Parlon S-20	Chlorinated rubber	Hercules
Pentalyn G	Pentaerythritol ester of rosin	Hercules
Plexigum N80	Resin	
Plexigum N742	Acrylic resin	
Polymin		
Polymist A-12		Allied
Post 4	Rheological additive	CasChem
Printex 35	Pigment	Degussa
Proxel GXL & BD	Biocides	Zeneca
QP-4400		UnCarbide
R49-HP-1210	Barium lithol presscake	Chromatic
Raven 450	Pigment	Columbian

RAW MATERIAL	CHEMICAL DESCRIPTION	SOURCE
Resimene U-980		
Rheolate 101	Thickener	Rheox
Rheolate 300	Thickener	Rheox
Rhoplex E-32	Acrylic binder	Rohm&Haas
Richgold Flake 9520X	Bronze powder	USBronze
S-232	Polyethylene wax	Shamrock
S-394-N1 Polyethylene	Wax (5 microns)	Shamrock
S-395-N5 Polyethylene	Wax (12.5 microns)	Shamrock
SAIB-90	Sucrose acetate isobutyrate	Eastman
Santicizer 141	Plasticizer	Monsanto
Santicizer 160	Plasticizer	Monsanto
SCX-631	Waterborne acrylic polymer	Johnson
SCX-660	Waterborne acrylic polymer	Johnson
Silane Z6020	Aminosilane derivative	DowCorning
Silberciser C6		
SST-3 PTFE Powder	PTFE wax	Shamrock
STAPA Hydroxyl W-F89	NL aluminum paste	ObronAtlan
Starfire Base Color	High-strength fluroescent pigment	DayGlo
Starfire SFB-22 Strong Magenta	fluorescent magenta	DayGlo
Sun Brite 273-3556	AAA yellow	Sun
Sun Brite Red 48:2	Calcium lithol red 2B (48:2)	Sun
Sun Brite Yellow 12	Diarylide yellow	Sun
Sun #U64-2357	Blue shade phthalocyanine green color	Sun

RAW MATERIAL	CHEMICAL DESCRIPTION	SOURCE
Sun Fast 234-0077	Pigment red	Sun
Sun Fast 248-1282	G.S. blue	Sun
Sun Fast Blue 15:3	Phthalo blue G.S.	Sun
Sun Fast 249-1291	Pigment blue	Sun
Sun Fast Blue 449-1281	Presscake (40% TSC)	Sun
Sun Fast Blue 449-5050	Phthalo G/S presscake	Sun
Sunsperse RHD 6017	Pigment dispersion	Sun
Super High Flash Naphtha		
Surcol 836W	Acrylic resin	
Surfynol DF-75	Defoamer	AirProduct
Surfynol OP-340	Additive	AirProduct
Surfynol 104E	Surfactant	AirProduct
Surfynol 104PA	Surfactant	AirProduct
Surfynol 104PG	Surfactant	AirProduct
Tamol 731	Surfactant	Rohm&Haas
Tecsol C Solvent	Denatured ethyl alcohol	Eastman
Tektamer 38LV	Biocide	Calgon
Tenneco 225F		Tenneco
Tilcon P1-2	Isopropoxyethoxy titanium bis acetyl-acetonate	
Tioxide R-TC 90	Titanium dioxide	
Ti-Pure R-960	Titanium dioxide	DuPont
Trionic 100	Solvent	Petrolite
TTDIA	Polymerized tung oil	Degen

RAW MATERIAL	CHEMICAL DESCRIPTION	SOURCE
Union Carbide LE410	Silicone food grade emulsion	UnCarbide
Unisperse Red FBN-PI	C.I. Red 2	Ciba-Geigy
Uvinul MS-40	Benzophenone-4	
Vehicle F	News flexo vehicle	Johnson
Vehicle H	News flexo vehicle	Johnson
Velex GQV-971	Gloss quickset vehicle	Day-Glo
Velex GQV-2030	Gloss quickset vehicle	Day-Glo
Velex MFV-2080	Matte fls. vehicle	Day-Glo
Velex TR-052	Tack reducer	Day-Glo
Vulkan K	Carbon black	Cabot
Wallkyd 11-151	Alkyd (35% in odorless M.S.)	Reichhold
WWX-1041	Wax	Carroll
XAMA-2	Crosslinker	
Zinc Oxide Solution #1	zinc/ammonium solution	Johnson
Zinplex 15	Crosslinker	UltraAdhes
70-MM-2230 Green Shade Phthalo Blue Flush		BASF

Section VI
Suppliers' Addresses

Air Products & Chemicals
Box 2662
Allentown, PA 18001
(215)-481-4911/(800)-345-3148

Allied-Signal Inc.
P.O. Box 2332
Morristown, NJ 07962
(201)-455-5010/(800)-526-0717

American Cyanamid Co.
Cytec Industries
5 Garret Mt Plaza
West Paterson, NJ 07424
(201)-357-3309/(800)-443-0443

Argus Div.
Witco Corp.
520 Madison Ave.
New York, NY 10022
(212)-605-3645/(800)-634-4010

BASF
100 Cherry Hill Rd.
Parsippany, NJ 07054
(201)-316-3939/(800)-426-8709

BYK-Chemie USA
524 South Cherry St.
P.O. Box 5670
Wallingford, CT 06492

Cabot Corp.
Cab-O-Sil Div.
Rte 36W
Tuscola, IL 61953
(217)-253-3370

Calgon Corp.
P.O. Box 717
Pittsburgh, PA 15230
(412)-787-6700/(800)-422-7266

Cas Chem Inc.
Cambrex Co.
40 Avenue A
Bayonne, NJ 07002
(201)-858-7900/(800)-227-2436

CDI Dispersions
27 Haynes Ave.
Newark, NJ 07114
(201)-824-1806

Columbian Chemicals Co.
1600 Parkwood Circle
Suite 400
Atlanta, GA 30339
(404)-951-5700/(800)-235-4003

Croda Inc.
7 Century Dr
Parsippany, NJ 07054
(201)-644-4900

DayGlo Color Corp.
4515 St. Clair Ave.
Cleveland, OH 44103
(216)-391-7070

Degen Co.
200 Kellogg St.
P.O. Box 5240
Jersey City, NJ 07305
(201)-432-1192

Degussa Corp.
65 Challenger Rd.
Ridgefield Park, NJ 07660
(201)-641-6100

Dow Chemical USA
Midland, MI 48674
(800)-447-4369

Dow Corning Corp.
Auburn, MI 48611
(517)-496-4000

Drew Industrial Div
Ashland Chemical Inc.
One Drew Plaza
Boonton, NJ 07005
(800)-526-1015

DuPont Co.
1007 Market St.
Wilmington, DE 19898
(800)-441-7515

Eastman Chemical Co.
P.O. Box 431
Kingsport, TN 37662
(615)-229-4511/(800)-EASTMAN

Exxon USA
P.O. Box 2180
Houston, TX 77252
(713)-656-3636/(800)-44-EXXON

Ferro Corp.
7050 Krick Rd.
Bedford, OH 44146
(216)-641-8580/(800)-321-9946

B.F. Goodrich Co.
9911 Brecksville Rd.
Cleveland, OH 44141
(216)-447-5000

Henkel Corp.
300 Brookside
Ambler, PA 19002
(215)-628-1476/(800)-531-0815

Hercules Inc.
Hercules Plaza
Wilmington, DE 19894
(800)-247-4372

J.M. Huber Corp.
907 Revolution St.
Havre de Grace, MD 21078
(301)-939-3500

Humko Chemical Div.
Witco Corp.
755 Crossover Lane-Ste 216
Memphis, TN 38117
(901)-684-7047

ICI Resins US
730 Main St.
Wilmington, MA 01887
(617)-658-6600/(800)-225-0947

S.C. Johnson & Son, Inc.
1525 Howe St.
Racine, WI 53403
(414)-631-3920/(800)-231-7868

Lawter International Inc.
990 Skokie Blvd.
Northbrook, IL 60062
(312)-498-4700

Magie Bros Oil Div.
Pennzoil Products Co.
9101 Fullerton Ave.
Franklin Park, IL 60131
(312)-625-2600

McIntyre Group Ltd.
1000 Governors Hwy
University Park, IL 60466
(708)-534-6200

Michelman, Inc.
9089 Shell Rd.
Cincinnati, OH 45236
(513)-793-7766

Mobay Corp.
Mobay Rd.
Pittsburgh, PA 15205
(412)-777-2000

Monsanto Co.
800 N. Lindbergh Blvd.
St. Louis, MO 63167
(314)-694-1000/(800)-325-4330

Mooney Chemicals Inc.
2301 Scranton Rd.
Cleveland, OH 44113
(216)-781-8383/(800)-321-9696

Morton International Inc.
150 Andover St.
Danvers, MA 01923
(978)-774-3100

Nalco Chemical Co.
One Nalco Ctr
Naperville, IL 60566
(312)-961-9500

Petrolite Corp.
6910 E 14 St.
Tulsa, OK 74112
(918)-836-1601

PMC Specialties Group
20525 Center Ridge Rd.
Rocky River, OH 44116
(216)-356-0700/(800)-543-2466

PPG Industries Inc.
3938 Porett Drive
Gurnee, IL 60031
(708)-244-3410/(800)-323-0856

Reichhold Chemicals, Inc.
525 N. Broadway
White Plains, NY 10603
(914)-682-5700/(800)-431-1920

Rheox Inc.
Wyckoff Mills Rd.
Hightstown, NJ 08520
(609)-443-2320

Rohm & Haas Co.
100 Independence Mall W
Philadelphia, PA 19106
(215)-592-3000

Ross Chemical Inc.
303 Dale Drive
Fountain Inn, SC 29644
(803)-862-4474

Shamrock Technologies Inc.
Foot of Pacific St.
Newark, NJ 07114
(201)-242-2999

Shepherd Chemical Co.
4900 Beech St.
Cincinnati, OH 45212
(513)-731-1110

Sun Chemical Co.
411 Sun Ave.
Cincinnati, OH 45232
(513)-681-5950/(800)-343-2583

Sutton Laboratories
116 Summit Ave.
Chatham, NJ 07928
(201)-635-1551

3M
3M Center
St. Paul, MN 55144
(612)-733-5454

Ultra Additives Inc.
460 Straight St.
Park Station, P.O. Box 98
Paterson, NJ 07543
(201)-279-1306/(800)-524-0055

Union Carbide Corp.
39 Old Ridgebury Rd.
Danbury, CT 06817
(203)-794-5300/(800)-342-3698

U.S. Aluminum Inc.
Route 202, P.O. Box 2190
Flemington, NJ 08822
(201)-782-5454

U.S. Bronze Powders, Inc.
Route 202, P.O. Box 31
Flemington, NJ 08822
(201)-782-5454

Witco Corp.
One American Lane
Greenwich, CT 06831
(203)-552-3382/(800)-634-4010

End Use Index

FSC
www.fsc.org
MIX
Papier | Fördert gute Waldnutzung
FSC® C083411

Kolibri 360 GmbH
Ferdinand-Jühlke-Straße 7
99095 Erfurt, Deutschland
produktsicherheit@kolibri360.de